Interdisciplinary Pedagogy for STEM

Reneta D. Lansiquot
Editor

Interdisciplinary Pedagogy for STEM

A Collaborative Case Study

Editor
Reneta D. Lansiquot
City University of New York
New York City College of Technology
Brooklyn, New York, USA

ISBN 978-1-137-56744-4   ISBN 978-1-137-56745-1   (eBook)
DOI 10.1057/978-1-137-56745-1

Library of Congress Control Number: 2016941867

Printed on acid-free paper

This Palgrave Macmillan imprint is published by Springer Nature
The registered company is Nature America Inc. New York

To my loving partner, mom, dad, and sister, a public school teacher. Also to educators who team-teach, sharing their multiple perspectives with students.

FOREWORD

Imagine a student earning a bachelor's degree in computer systems at your institution, going on to earn a PhD in educational communication and technology, returning to your institution as an assistant professor in the English Department, and within a few years, leading an initiative to expand interdisciplinary teaching and learning, which becomes a signature component of the college's new General Education program. Such was the trajectory of Dr. Reneta D. Lansiquot, editor of this book, who first approached me when I was serving as the dean of the School of Arts and Sciences with the idea of forming an Interdisciplinary Studies Committee to promote interdisciplinary curricular development. The introductory chapter, by Professor Reneta D. Lansiquot, traces her story, focusing on her experiences both as a student and as a faculty member. It also includes practical information on administrative issues, pedagogical approaches, and navigating institutional politics for those inspired to lead a similar initiative at their own institution.

A college's General Education program is that portion of the curriculum that all students share. It helps to define an institution's identity and represents a consensus on what students should know, value, and be able to do. The remaining chapters in this book demonstrate the depth and breadth of the interdisciplinary courses developed as part of this initiative and how they promote student's critical thinking.

In Chap. 2, Philosophy Professor Laureen Park describes how students integrated various disciplinary perspectives in the interdisciplinary course, *Weird Science: Interpreting and Redefining Humanity*, using Hans Georg

Gadamer's notion of the *sensus communis*. Ways that students build on prior knowledge to integrate interdisciplinary concepts are highlighted.

Economics Professor Sean P. MacDonald and Sociology Professor Costas Panayotakis analyze the concept of an insatiable human nature through a critique of traditional neoclassical economic assumptions about human behavior in Chap. 3. Through a case study on the causes of the global economic and financial crisis of 2008, MacDonald and Panayotakis argue that actions traditionally accepted by many as "natural" are actually cultivated social mores. This activity challenges students to question assumptions that are presented as truths.

Chapter 4, by Library Science Professor Anne E. Leonard and Psychology Professor Jean E. Hillstrom, describes the development and content of the responsible conduct of research and information literacy modules, as well as discussing the role of each in an interdisciplinary course. Leonard and Hillstrom make a clear connection between interdisciplinary student learning outcomes related to skills, values, and constructivist theory.

Professor Reneta D. Lansiquot and Computer Systems Technology Professor Candido Cabo describe the design, development, and teaching of an interdisciplinary course linking creative writing and computational thinking for non-computer majors in Chap. 5. In this course, students create original stories and then implement them as a video game prototype using computer programming.

Chapter 6, written by Physics Professor Reginald A. Blake and Mathematics Professor Janet Liou-Mark, presents research findings related to their work as principal investigators on National Science Foundation Research Experiences for Undergraduates (NSF REU) grants in the geosciences. Through undergraduate research, students actively and collaboratively discover new knowledge while learning more about one of the most interdisciplinary scientific fields, the geosciences. Results indicate that geoscience research experiences increased the students' understanding of the relevancy of their interdisciplinary study to society.

In the last chapter, Professor Reneta D. Lansiquot and graduate student Tamrah D. Cunningham present a student's perspectives of the unique challenges and distinct advantages of team-taught interdisciplinary courses.

In summary, this book provides rich food for thought for educators, while providing a road map to those interested in adopting such a productive approach at their own institution. New York City College of Technology is proud of the accomplishments of Professor Lansiquot in her roles as both an alumni and faculty member.

Associate Provost Pamela Brown
New York City College of Technology
Brooklyn, NY

Contents

Notes on Contributors

Reneta D. Lansiquot is Associate Professor and Program Director of the Bachelor of Science in Professional and Technical Writing at New York City College of Technology of the City University of New York, where she earned an AAS in computer information systems and a BTech in computer systems. She earned an MS in integrated digital media at Polytechnic University and a PhD in educational communication and technology at New York University. Her research focuses on interdisciplinary studies. Her first book is entitled *Cases on Interdisciplinary Research Trends in Science, Technology, Engineering, and Mathematics: Studies on Urban Classrooms.* This new book, *Interdisciplinary Pedagogy for STEM: A Collaborative Case Study,* is being released simultaneously with *Technology, Theory, and Practice in Interdisciplinary Pedagogy for STEM: A Collaborative Study and Non-STEM Approaches.*

Reginald A. Blake is Associate Professor of Physics at New York City College of Technology of the City University of New York. He is a geophysicist who serves on the New York City Panel on Climate Change and as a NOAA-CREST scientist. He is a member of the Climate Change Impacts team at NASA-GISS, Columbia University, and he directs the Center for Remote Sensing and Earth System Sciences. Dr. Blake has secured grant funding as Principal Investigator or Co-PI from the NSF (REU, OEDG, IUSE), NOAA, NASA, and DOD. He is the director of the Black Male Initiative Program, and he has published and presented extensively on climate change impacts, on satellite and ground-based remote sensing, and on promoting STEM education for underrepresented minority students.

Pamela Brown is Associate Provost at New York City College of Technology of the City University of New York. She earned a PhD in chemical engineering from Polytechnic University, an MS in chemical engineering practice from MIT, and a BS in chemistry from the University at Albany, SUNY. Dr. Brown has published in journals ranging from *AIChE* to *JCE*. The principal investigator or Co-PI on five

NSF grants totaling over $3 million, she has also served as program director at the National Science Foundation's Division of Undergraduate Education. She currently serves on the National Research Council of the National Academy of Sciences Committee, "Barriers and Opportunities in Completing Two and Four Year STEM Degrees," which is preparing a consensus paper for early 2016 release.

Candido Cabo is Professor of Computer Systems Technology at New York City College of Technology of the City University of New York and a member of the doctoral faculty at the CUNY Graduate Center. He earned the degree of Ingeniero Superior de Telecomunicacion from the Universidad Politecnica de Madrid, and a PhD in Biomedical Engineering from Duke University. He was a research scientist in the Department of Pharmacology at the College of Physicians and Surgeons of Columbia University. He has written a number of peer-reviewed journal articles on computer modeling of biological processes, particularly relating to cardiac electrophysiology. His research interests include computer science education and the use of computational models to understand and solve problems in biology.

Tamrah D. Cunningham is an adjunct professor at New York City College of Technology of the City University of New York where she earned a BTech in computer systems. She is currently completing her MFA in game design at New York University. Her research interests include narrative studies and world-building in role-playing games, particularly Japanese RPGs. She has designed several games, most recently, *Everlasting Unemployment*, a choice-based adventure game parodying the struggles of finding employment after graduating from college. She is currently working on a mobile game that will educate its players on chronic child illness.

Jean E. Hillstrom is the Chair of the Social Science Department at New York City College of Technology of the City University of New York, and she has held several academic appointments including associate professor at Adams State College Colorado. She earned her PhD in Applied Cognitive Aging/Developmental Psychology at The University of Akron. Her research interests are varied and include topics such as adjudicative competency in juveniles; emotion regulation; trauma, stress, physiological arousal, and coping; age and job satisfaction; and age-and job-related training performance. She has presented at local, national, and international conferences and has also published in scholarly journals. Dr. Hillstrom has extensive experience with human subjects research protections and is currently vice chair of the CUNY Integrated IRBs.

Anne E. Leonard is the coordinator of Information Literacy and Library Instruction at New York City College of Technology of the City University of New York. She earned an MLIS from the University of Texas at Austin and an MS in Urban Affairs from Hunter College. In addition to teaching "Research and Documentation for the Information Age," she teaches in the library's instruction program, offering research and information literacy instruction to classes in English and civil engineering technology. Her academic interests include the professional status of academic librarians, critical information literacy, and place-based learning in undergraduate education. Her interests in cities and information literacy merged in the classroom when she co-taught a special topics interdisciplinary course, "Learning Places: Understanding the City," with Architectural Technology faculty.

Janet Liou-Mark is Professor of Mathematics and Director of the Honors Scholars Program at New York City College of Technology of the City University of New York. Her research focuses on the Peer-Led Team Learning instructional model has won her the 2011 CUNY Chancellor's Award for Excellence in Undergraduate Mathematics Instruction. Dr. Liou-Mark is the recipient of several federal and foundation grants. She is currently a co-principal investigator of the Math Science Partnership (MSP), Research Experience for Undergraduate (REU), and Improving Undergraduate STEM Education (IUSE): Pathways into Geoscience NSF grants. Dr. Liou-Mark has mentored over 125 underrepresented minority students; a third are continuing or obtaining advanced degrees. She organizes and speaks at women conferences in Malawi, Africa, and builds libraries there.

Sean P. MacDonald is Associate Professor of Economics at New York City College of Technology of the City University of New York. She earned a BA in Sociology from the University of Maryland and a PhD in Economics from the New School for Social Research in New York. She has published several papers on the 2008 housing and financial crisis. Following creation of a new interdisciplinary course, "Environmental Economics," her research interests include collaborative interdisciplinary projects, and the benefits of place-based undergraduate research. She continues research and writing on the longer-term impact of the financial crisis on worsening income inequality.

Costas Panayotakis is Professor of Sociology at New York City College of Technology of the City University of New York. After earning a BA in Economics from Stanford University, he received his PhD in Sociology from CUNY's Graduate Center. He is the author of *Remaking Scarcity: From Capitalist Inefficiency to Economic Democracy*. In addition to his scholarly work, which includes numerous articles in economics, philosophy, and sociology journals, he has been interviewed by dozens of media outlets around the world, including the BBC, Al Jazeera, and RT International, and interviewed on programs such as *Democracy Now!*.

Laureen Park is Associate Professor of Philosophy at New York City College of Technology of the City University of New York. She earned her PhD in Philosophy from the New School for Social Research. Her research interest involves phenomenological methods in analyzing empirical problems, as she does in "Opening the Black Box," in the *International Journal for Peace Studies*, an interdisciplinary journal. Recent activities furthering her interest in interdisciplinary studies include developing an interdisciplinary logic course that integrates mathematical and philosophical perspectives.

LIST OF TABLES

CHAPTER 1

Introduction: An Interdisciplinary Approach to Problem Solving

Reneta D. Lansiquot

Abstract This introductory chapter traces my story. It focuses on my experiences as a student and faculty member at New York City College of Technology, which is the designated college of technology of the City University of New York. It chronicles the creation of interdisciplinary courses and research experiences that help students tackle complex problems and make connections between their major and general education courses. Practical information at the administrative level, such as governance issues, developing an application process for interdisciplinary course designation, forming an interdisciplinary studies committee, and the responsibilities of such a committee, are discussed. This chapter provides pedagogical strategies for team-teaching, as well as highlights supportive tools such as original case studies written by faculty for interdisciplinary courses.

Keywords College governance • Interdisciplinary studies • Pedagogical case studies • Team-teaching

R.D. Lansiquot (rlansiquot@citytech.cuny.edu ✉)
Department of English, New York City College of Technology,
City University of New York, Brooklyn, NY, USA

© The Editor(s) (if applicable) and The Author(s) 2016
R.D. Lansiquot (ed.), *Interdisciplinary Pedagogy for STEM*,
DOI 10.1057/978-1-137-56745-1_1

1

Although interdisciplinary pedagogical strategies focus on problem-based learning, they are not limited to it, as they nurture a host of other valuable skills and attitudes. As a result of interdisciplinary experiences, students learn to distinguish the perspectives of different disciplines, purposefully connecting and integrating knowledge and skills from across disciplines to solve problems. As students synthesize and transfer knowledge across disciplinary boundaries, they become flexible thinkers who are comfortable with complexity and uncertainty, and they learn to understand the factors inherent in complex problems as well as grasping the universal nature and deep structure of science. This approach to problem-solving prepares students for their future as lifelong learners, while encouraging them to apply their capacity as integrative thinkers to solve problems in ethically and socially responsible ways.[1] Interdisciplinary studies challenge learners uniquely to think critically, communicate effectively, and work collaboratively with others.

With the breadth of expectations concerning interdisciplinary learning in mind, this book connects constructivist theory and practice. It examines the successful collaborative interdisciplinary studies at New York City College of Technology (City Tech), which is the designated college of technology of the City University of New York (CUNY). City Tech is dedicated to increasing participation, retention, and graduation rates in science, technology, engineering, and mathematics (STEM) academic disciplines in an institution that is known for its extremely diverse student body, and places particular emphasis on students underrepresented in those fields. The implementation of constructivist theory complements interdisciplinary studies, as constructivist learning includes embedding learning in complex, realistic, and relevant settings. Constructivist learning promotes social negotiation, an integral part of learning, while supporting the consideration of multiple perspectives and the use of multiple modes of representation. As a result, students take ownership of their learning and gain self-awareness of the knowledge construction process.[2] This emphasis on theory is complemented by the specificity of experiences presented by the chapter authors of this book.

I am uniquely aware of the benefits of social negotiation in learning. A little over a decade ago, I was a student at City Tech in the Computer Systems Technology (CST) department, whose studies focused on computer programming. There was an average of three other female students in my classes. A couple of semesters into the two-and-a-half-year period that I was as an undergraduate, I enrolled in a learning community. This learning community merged a web programming course with a graphic arts course. As a student in these courses, I programmed the "back end" of a webpage—the computer code, connections between database and servers, and so forth—and then went to the next class to learn how to design an aesthetically pleasing "front end"—that is, the part the user can see, such as drop-down menus. I am not sure how I managed to enroll in this learning community, but I benefitted immensely from the experience of learning how the different perspectives on the same task could be used to create a well-designed, functional website.

My experiences in this interdisciplinary learning community were foundational. Years later, after studying integrated digital media, educational communication, and technology in graduate school, I returned to City Tech as a tenure-track assistant professor. During my first semester, I gave the keynote address at the induction ceremony of the National Society of Collegiate Scholars at the college, a chapter that I charted as a student. The title of the speech was "On the Need for Interdisciplinary Studies." I spoke of my own interdisciplinary background, encouraging students to enroll in a few courses outside their major that piqued their interests. Although I was a computer programmer, I was drawn to English courses like Perspectives in Literature, as well as Mathematics and Physics courses. However, in reality, being able to enroll in such courses was not possible for most students. As City Tech is a STEM-focused institution that prepares students for careers in industry, most students who are enrolled in their desired majors begin taking courses in their major right way, and there is little, if any, room for electives. This meant that if a faculty member created an innovative interdisciplinary course, it would likely not gain the necessary enrollment to be offered (unless it was required as part of a degree program) because it would be a non-contributory fall-through course that financial aid would not cover. This is a significant hindrance, because the majority of students at the college depend on financial aid to pay for all or part of their education.

Notwithstanding, it was clear to me that creating interdisciplinary courses at City Tech would provide students with the opportunity to make connections between their major and general education courses. More importantly, these courses fostered the development of creative problem-solving. To do this, it was necessary to have practical information at the administrative level,[3] such as forming an interdisciplinary studies committee, the responsibilities of such a committee, governance issues, and developing an application process for interdisciplinary course designation.

Equally important were pedagogical strategies for teams of teachers, as well as supportive tools, such as open-source technology and original case studies written by faculty to support interdisciplinary courses.[4]

BUILDING THE FOUNDATION FOR INTERDISCIPLINARY STUDIES

Shifting from my experience as a student frustrated by the apparent lack of connections between my required liberal arts courses and my technology career ambitions to my development into one who valued and had expertise in interdisciplinary studies, I took advantage of an opportunity to become involved in the Project Kaleidoscope (PKAL) Facilitating Interdisciplinary Learning (FIDL) project, which began in 2007. I was a member of one of the 28 campus teams engaged in sharing the experience of building and sustaining interdisciplinary programs in science and mathematics.[5] I traveled around the country, participating in STEM-focused interdisciplinary workshops and think tanks. At the end of my first year as a faculty at City Tech, I was accepted to and completed the PKAL Summer Leadership Institute for STEM Faculty. This led me to taking on the charge by the founding director of PKAL, Jeanne L. Narum, to become a "positive deviant" on campus.

At the start of my second year, in fall 2009, I asked the dean of the School of Arts & Sciences, Pamela Brown, to create an Interdisciplinary Studies Committee and add it to the list of existing committees (all faculty of the school are placed on one committee; faculty can request to change committees). This committee was charged with pursuing opportunities to develop and help sustain interdisciplinary curricula. We researched the state of interdisciplinary courses and the use of the word "interdisciplinary" at all colleges within CUNY.[6] That same year, the college received a million-dollar, five-year grant from the National Science Foundation (NSF), *The City Tech I³ (Innovation through Institutional Integration) Incubator: Interdisciplinary Partnerships for Laboratory Integration.* A year later, in fall 2010, we developed and administered an Institutional Review Board-approved faculty survey to track and record a baseline of interdisciplinary activities at City Tech. We used this survey to determine what types of workshops we should offer faculty. We co-sponsored a workshop series on creating, implementing, and assessing interdisciplinary STEM projects. By 2012, my role as a Co-Principal Investigator on the grant provided me the opportunity to call for the Interdisciplinary Studies Committee to become a college-wide Interdisciplinary Committee. As this was a new committee, no mechanisms were in place, and everything had to be done from scratch. I had to create policies with the help of committee members.

During this time, CUNY was in the process of revising the general education program, now requiring all colleges in the university system to have the same general education requirements. However, due to the unique nature of the colleges, it became necessary to include a "College Option" that consisted of three course requirements of each college's choice. As a good number of members of the Interdisciplinary Committee were also on the college-wide General Education Committee at City Tech, we were able to push for the inclusion of an interdisciplinary course. Due, in part, to my efforts, as of fall 2013, all students graduating with a bachelor's degree are required to take a team-taught interdisciplinary liberal arts and sciences course as part of their general education at the College.

After the College Option was approved, on November 5, 2013, it became necessary to promote an interdisciplinary campus culture and to formalize the process for interdisciplinary course designation, ensuring appropriate curriculum development and governance. At the time, the interdisciplinary course approval process was very complex, requiring multiple submissions to the same committee. The College's Interdisciplinary Committee is charged with helping faculty design and implement interdisciplinary pedagogy. In addition, it is a recommending body to the College Council's Curriculum Committee (in consultation with the Provost and the Associate Provost). As part of a comprehensive major curriculum change proposal designed to, in part, streamline the process for the submission of new course proposals, the Interdisciplinary Committee created relevant documents: "Criteria for an Interdisciplinary Course," which includes the definition of interdisciplinary studies; "Application for Interdisciplinary Course Designation;" "Submitting an Interdisciplinary Course Proposal," which described the process; and "Suggestions for Reimagining a Course as Interdisciplinary." We also provided a list of four new Interdisciplinary Committee subcommittees (New Course, Existing Courses, Course Development, and Social Outreach), along with their duties.

After creating the mechanisms for interdisciplinary courses, our role still had to be approved by College Council. Gaining support was a difficult, hard-fought battle. For example, a large number of the Mathematics faculty and some members of the English Department, voting members of College Council, vowed to oppose this proposal. There was an Open Hearing to allow faculty the chance to ask questions or voice concerns. With the exception of one Mathematics faculty member—who was the then-chair of the College Council's Curriculum Committee and was required to be in attendance—no other Mathematics faculty attended. Further along in the process, a Mathematics faculty member deliberately stalled the process by not placing the proposal on the agenda to be voted on by the entire Council (this can only be done once). In addition, members of the College Council's Executive Committee, which is composed of all the chairs of the standing committees (at the time, roughly half were Mathematics faculty), were led to believe that ours was a new committee with unknown membership and little background at the college. For this reason, when the proposal was finally put on the agenda, I included the names of all 21 committee members to show that the proposal had considerable support throughout the college. When those who were opposed to the proposal were challenged to state the reasons for their opposition, it became clear they had not read it. I then took a different tack by simply explaining the rationale for the main issue of contention—namely, the team-teaching requirement and pointing out that no faculty member is required to teach an interdisciplinary course.

In this proposal, we had to provide justification for the need for interdisciplinary courses[7] and define interdisciplinary courses for City Tech. We characterized the courses in this way:

> Although many academic disciplines, such as African-American Studies and Engineering, are inherently interdisciplinary, to be considered an interdisciplinary course at City Tech the course must be team-taught[8] by more than one faculty member from two or more departments in the College. An interdisciplinary course, by definition, has an interdisciplinary theme as its nucleus. In its essence, such a course brings the analytic methods of two or more academic disciplines to bear on a specific problem or question. Thus, a course in Music History is not likely to be considered interdisciplinary, but a course in Music History from an economist's perspective might very well lead to such a course. The application of different methods and concepts is the key to assessing whether a course is or is not interdisciplinary. The term, interdisciplinary, is occasionally used to identify individual projects or assignments, but these, though possibly commendable, fall short in the necessary scope for learning experiences that demand in-depth exposure to the methodologies of distinct intellectual disciplines, and the creative application of these methodologies to specific problems.

We decided on the team-teaching requirement to facilitate collaboration and to help ensure that multiple perspectives would be represented during the course (after all, I have studied several disciplines, but I bring my one perspective into the classroom). Interdisciplinary team-taught courses involve two or more educators from different disciplines actively sharing the instruction and evaluation of the content and skills in the same classroom to provide students with multiple perspectives and to facilitate learning in interdisciplinary courses. We also provided suggestions for reframing a course as interdisciplinary in the proposal. The following is a list of some of the ways that a course could be organized in order to fulfill interdisciplinary requirements.

- *Shared credits*: Two faculty split the credits of the same course. Faculty will not receive double workload hours for the same course.[9] For example, a course with three contact hours equates to a total of three workload hours that are appropriately split among teaching faculty (there may be more than two faculty). Although a one-credit lab will not meet the criteria, it could be part of a four-credit science interdisciplinary course. If a three-credit course deems six workload hours for two faculty, then this course should be reconfigured so the three workload hours can be split amicably and fairly.
- *Trading credits*: Two faculty agree to teach two separate sections of courses, both interdisciplinary, but not necessarily the same course. They trade lessons between the two courses, so that workload is equalized, but both courses gain the interdisciplinary designation.
- *Guest lecturers*: The course provides multiple perspectives via experts who deliver the interdisciplinary content. This may include, but should not be limited to, site visits.
- *Learning community*: Two courses are assigned as a learning community, and two faculty provide the divergent viewpoints. One course becomes officially labeled with the interdisciplinary designation, and the other course is assigned credit to some other area in the student plan of study.

The college's current learning communities do not satisfy the General Education interdisciplinary course requirement for baccalaureate students because they are not upper-level courses with at least one prerequisite. Also, these first-year linked courses include a cohort of students who enroll in the same two or three courses in different disciplines that are, at minimum, connected by a theme or, at most, include a common assignment.

- *Independent study*: A learning experience that allows for self-directed study relating to an area of academic or professional experience. Key elements of the course include critical analysis, application or development of ideas and concepts related to the area of inquiry, and guidance by two or more faculty mentors.

Faculty teaching interdisciplinary courses are encouraged to consider using free/open-access course materials whenever possible to avoid requiring students to buy double books. The proposal also noted that as an interdisciplinary course will probably be new to students, background information should be included in course syllabi, including a clear explanation of the scheduling of participating faculty and needed contact hours. To facilitate scheduling, an interdisciplinary course designed to be offered fully or partially online should be considered, as appropriate, and is encouraged.[10]

Once the proposal was approved, its membership had to be sorted out. I drafted the membership policy based on my experience working on other committees to avoid their pitfalls. After incorporating feedback from the members to clarify policies, we approved it on December 12, 2013. Members were now to be drawn from all three schools at the college: Arts & Sciences, Technology & Design, and Professional Studies. This policy states that the committee will be composed of faculty members interested in the development of interdisciplinary courses at City Tech. For logistical purposes, membership in the committee is capped at 24 voting members and 8 non-voting members. There are to be no more than two members of the committee from the same department (the founding members were noted on this one-page document). If a department already has two members on the committee, then additional faculty from that department can only serve as non-voting members. If more than two committee meetings are missed in a given semester without an adequate reason (e.g., the faculty member is teaching, at a conference, sick, or must attend an Appointments or Ad Hoc Promotion Committees meeting), the faculty will be reconsidered for membership on the committee.[11]

In order to maintain the cross-disciplinary nature of the committee, faculty interested in becoming a voting member of the committee can do so if the number of committee members is below 24 and if there are fewer than two members from their department in the committee. If those circumstances do not apply, the faculty member can join the committee as a non-voting member. Non-voting members can become voting members when there is a voting-member opening in the committee. There are 24 committee voting members. Additional members are considered as non-voting members. To ensure that all full-time faculty at City Tech are given an opportunity to participate in the committee, we implemented a term limit of three years. Every year, no more than one-third of the committee will be replaced to ensure that at least two-thirds of the committee are experienced members. A committee member who has already served for three years may reapply after one year. To ensure continuity, the aforementioned existing members can serve three years from the semester in which this membership policy was approved, fall 2013, before the term limits start. At the end of these three years, one-third of these members will need to retire to start the cycle.

If membership falls below 24, with no additions, then current members may continue to serve. Faculty interested in serving alert the chair of the committee. If there are no available positions, then their name will go onto a waiting list. When a position opens, the first member on the list is evaluated; if there is an opening from that department, then he or she is offered the position. If an individual does not accept, he or she is removed from the list and the next in line is offered, until the positions are filled. In another provision to ensure continuity in the committee, a three-fifths majority of the voting members is needed to make amendments to the foundational rules.

These rules were formulated to promote and facilitate interdisciplinary courses like Weird Science: Interpreting and Redefining Humanity.

Writing, Implementing, and Assessing Case Studies

Case studies are a pedagogical strategy that can be used as interdisciplinary teaching tools. An effective case study tells a story, focuses on an interest-arousing issue, creates empathy with the central character, is relevant to the reader, and is short.[12] Writing and implementing case studies—as well as constructing a good team of teachers and using team-teaching models such as traditional, collaborative, complementary or supportive, and monitoring teachers—promote interdisciplinary studies in the classroom.

Before offering my new writing-intensive interdisciplinary course, Weird Science: Interpreting and Redefining Humanity, I asked guest lecturers to write case studies that could serve as modules encapsulating their perspectives on the question of the course: What does it mean to be (virtually) human?[13] The course introduces different disciplines through perspectives on this enduring question. It is meant to excite and engage students in communicating this complex question using digital media and to encourage underrepresented first-year student groups to consider various disciplines. This innovative course, which is housed in the English Department, now hosts a dozen guest lecturers throughout the semester (from a diverse array of disciplines, including Philosophy, Biology, Psychology, Economics, Sociology, Library Science, Computer Science, Entertainment Technology, Architectural Technology, Mathematics, and Physics) providing perspectives on what it means to be human from perspectives ranging from that of molecular biology and quantum physics to that of economics and sociology, information literacy, and entertainment technology. I collaborate with faculty participating in this course before, during, and after the class to ensure that students are able to make clear connections (e.g., a different idea of rationality is presented by the philosopher and the economist) to the different lectures and assess student learning based on stated learning outcomes.

In order to help students make connections among disciplines and modes of thought, different models of team-teaching are used throughout this course. Traditional team-teaching is the guest lecturer, and I actively share the instruction of the content and skills to all students. For example, I may present the new material to the students while the guest lecturer constructs a concept map as the students listen to me. Adhering to collaborative team-teaching, there are student-led discussions and small-group work. Guest lecturers also discuss and exchange ideas in front of the students (e.g., during a shared guest lecture with an economist and a sociologist, as well as one with an architect and a mathematician). Using complementary or supportive team-teaching, I am responsible for teaching the content to the students, which a librarian, for example, takes charge of providing follow-up activities on information literacy skills. During guest lecturer activities, I often assume the role of a monitoring teacher, circulating the classroom and monitoring student understanding.

The experiences teaching this course have inspired this book. All contributors wrote case studies for this course and were part of this course from its first iteration five years ago. Writing as problem-solving—through case studies, annotated bibliography, literature reviews, and term papers—is a critical aspect of interdisciplinary studies. This interdisciplinary team-taught course, now offered as hybrid (i.e., taught mainly in a computer lab classroom and partially online on days that the class meets for half the time; extended classes as necessary to accommodate guest lectures and to facilitate group work) and traditionally, serves as a through line for the first three chapters of this book, connecting their original case studies discussed in conjunction with their presentations as guest lecturers.

SUSTAINING SPECIAL TOPICS INTERDISCIPLINARY COURSES

As the function of the Interdisciplinary Committee and the process for interdisciplinary course designation proposal was approved at the same time an interdisciplinary course was required of students, we needed inter-disciplinary courses, quickly. The new course approval process, although streamlined in recent years, is slow (taking, at best, a full academic year to be approved if there are no major revisions). As it already fit City Tech's criteria for an interdisciplinary course, Weird Science was the first course to be designated as interdisciplinary. As I worked on developing a new programming narratives[14] course, it became clear that we needed to reframe some existing courses as interdisciplinary. This meant course sections could now be interdisciplinary; that is, a Language and Thinking course that is housed in the English Department could be co-taught with a psychologist and gain interdisciplinary designation, while a different sec-tion of this course could be taught solely by an applied linguist. We also needed to clarify the mapping of the courses to the professions for stu-dents in majors; that is, specialized courses that focused on the profession are not liberal arts and science courses, and therefore do not meet the General Education "College Option" for an interdisciplinary course and were actually inter-professional, not interdisciplinary.

Our advocacy for interdisciplinary studies led to the inclusion of inter-disciplinary thinking as a Focus Goal in the college's 2014–2019 Strategic Plan. As the founding chair of this committee, I have been spearheading the development of exciting new interdisciplinary courses and promoting an interdisciplinary campus culture through information sessions, requests for new course proposals, the administration of mini-grants, and organiz-ing workshops, lecture series, and other campus-wide events. Through its outreach efforts and advocacy, our committee has already surpassed the goal of "at least 10 new and restructured" interdisciplinary courses that the college plan set for us. However, we needed 20 course sections in the near future to ensure that all students who need to graduate have an inter-disciplinary course available.

In order to expand the number of interdisciplinary courses available to students, in May 2014, *The City Tech I³ Incubator*[15] offered summer salary to groups of two or more faculty members from different disciplines to design new STEM-focused special topics interdisciplinary courses. This request for proposals resulted in three courses with a STEM focus: Energy Resources, an environmental science course housed in the Chemistry Department and proposed by a Chemistry and a Construction Management and Civil Engineering Technology professor; The Learning Places: Understanding the City, cross-listed in the Architectural Technology Department and Library, and proposed by faculty from these departments; Science in the Kitchen, housed in the Physics Department and proposed by a Physics and a Hospitality Management professor. In addition, two other courses were designed that did not qualify for NSF support because they were not STEM focused, but they were ultimately supported by the Office of the Provost: Healing the Body: The Visual Culture of Medicine, housed in the Humanities department and proposed by faculty from Art History, Nursing, and Dental Hygiene; and The Evolving Face of Race, Class, and Gender Identity, housed in the Human Services department and proposed by the chair of this department and English faculty. A written report summarizing implementation and lessons learned was required after the course was team-taught the first time.

These new special topic interdisciplinary team-taught courses provide a shell for faculty who did not propose these courses to teach. For example, a Theater professor from the Humanities Department is interested in co-teaching the special topics interdisciplinary course, Learning Places, with an Architectural Technology professor, and Science in the Kitchen will be initially taught by the proposers, a physicist and a chef, but other sciences may join the "kitchen," such as Biology, Chemistry, or even Anthropology and Economics.

Most recently, in May 2015, the committee, with the support from the Office of the Provost, offered small stipends for faculty to reframe existing courses, resulting in three approved courses: History of Theatre: Technology and Stages, housed in the Humanities Department and co-taught by a Theater professor and an Architectural Technology professor; Science Fiction, housed in the English Department and co-taught by an English professor and a Physics professor; and Health Care Ethics, housed in the Social Sciences Department and team-taught by a Philosopher professor with guest lecturers from Biology, Nursing, and Dental Hygiene. Currently, we have 18 courses designated as ID (for half of these courses, all sections will always be designated as interdisciplinary because these are new courses conceived to be interdisciplinary, not existing courses reframed to meet the college's interdisciplinary requirement), and there are several additional sections.[16] For the first time next semester, we will meet, and likely exceed, the 20 course sections needed to ensure that a General Education ID course is available to all students who need it.

Although interdisciplinary studies are not new to the college, there is still work to be done. Currently, sustaining interdisciplinary courses is a precarious endeavor, as there is no overarching structure, such as a Department of Interdisciplinary Studies or an Interdisciplinary Studies Program with a program director. There is also the need for simplified scheduling because interdisciplinary courses are housed in several departments, including those outside the School of Liberal Arts & Sciences. Also, a designated code for interdisciplinary courses (i.e., IDS 2000 course code or an ID section code to be in line with the college's current section codes such as D for day, E for evening, and HD for hybrid—that is, partially online—courses) would curb student confusion when registering for an interdisciplinary section of an existing course. Further, updated student evaluation of teaching and faculty classroom observation forms are needed in order to address team-teaching practices.

Whereas the need for interdisciplinary courses is well established, equipping and convincing the high numbers of contingent faculty, as well as probationary faculty with mounting pressures to receive tenure, to engage in team-teaching such courses remains difficult. Even though contemporary educational theory encourages team-based learning, integrating faculty (who are accustomed to having control over their syllabi, teaching methods, classroom management, and grading policies) into seamless, collaborative interdisciplinary teams is a challenging task. However, the value of offering students multiple perspectives from different faculty, helping them to make connections across disciplines, and fostering interdisciplinary educational research and an interdisciplinary campus culture justifies the effort.

The latest COACHE survey of Faculty Satisfaction[17] that was administered throughout CUNY last spring found "very positive responses to opportunities for collaborative work at the college and promising responses to interdisciplinary work. This will be hugely important as we move to develop more interdisciplinary opportunities and programs."[18] The contributors to this book, *Interdisciplinary Pedagogy for STEM: A Collaborative Case Study,* evaluate my Weird Science interdisciplinary course from a variety of different disciplinary perspectives: English, Philosophy, Library Science, Psychology, Economics, Sociology, and

Computer Science. The book includes a discussion of undergraduate research (specifically focused on Geosciences, Mathematics, and Physics). This book provides successful examples of interdisciplinary studies to inspire implementation on your campus, in your classroom, and for place-based learning.

NOTES

1. Project Kaleidoscope, *What Works in Facilitating Interdisciplinary Learning in Science and Mathematics: Summary Report* (Washington, DC: AAC&U, 2011); Reneta D. Lansiquot, *Cases on Interdisciplinary Research Trends in Science, Technology, Engineering, and Mathematics: Studies on Urban Classrooms* (New York: Information Science Reference, 2013).

2. Jean Lave and Etienne Wenger, *Situated Learning: Legitimate Peripheral Participation* (Cambridge: Cambridge University Press, 1991), and the zone of proximal development (ZDP) as discussed in Lev S. Vygotsky, *Mind in Society: The Development of Higher Psychological Processes* (Cambridge, MA: Harvard University Press, 1978). Also refer to David R. Krathwohl, "A Revision of Bloom's Taxonomy: An Overview," *Theory into Practice* 41, no. 4 (2002): 212–218; Benjamin S. Bloom, Max B. Englehart, Edward J. Furst, Walter H. Hill, and David R. Krathwohl, *Taxonomy of Educational Objectives, the Classification of Educational Goals, Handbook I: Cognitive Domain*, ed. Benjamin S. Bloom (New York: McKay, 1956); Rand J. Spiro, Paul J. Feltovich, Michael J. Jacobson, and Richard L. Coulson, "Cognitive Flexibility, Constructivism, and Hypertext: Random Access Instruction for Advanced Knowledge Acquisition in Ill-Structured Domains," in *Constructivism and the Technology of Instruction: A Conversation*, eds. Thomas M. Duffy and David H. Jonassen (Hillsdale, NJ: Lawrence Erlbaum Associates, 1992), 57–76; Jean Piaget, *To Understand Is To Invent: The Future of Education* (New York: Grossman, 1973).

3. For their human resources, structural, political, and symbolic four frames approach to tackling the challenges of sustaining interdisciplinary programs, see Susan Elrod and Mary J. S. Roth, "Framing Leadership for Sustainable Interdisciplinary Programs," *Peer Review* 17, no. 2 (2015): 8–12. Also refer to Susan Elrod and Mary J. S. Roth, *Leadership for Interdisciplinary Learning: A Practical Guide to Mobilizing, Implementing, and Sustaining Campus Efforts* (Washington, DC: AAC&U, 2012).

4. Cross-fertilization, team-collaboration, field creation, and problem orientation are categories of interdisciplinarity offered by Diana Rhoten and

Stephanie Pfirman, "Women, Science and Interdisciplinary Ways of Working," *Inside Higher Ed*, October 22, 2007. https://www.insidehighered.com/views/2007/10/22/rhoten

5. See Project Kaleidoscope, *What Works in Facilitating Interdisciplinary Learning in Science and Mathematics: Summary Report.*

6. More detailed information can be found in the chapter, "Theories into Practice: A Focus on STEM at City Tech," written by committee members Sean P. MacDonald, Olufemi Sodeinde, and Andleeb Zameer for the section on the college in my first book, *Cases on Interdisciplinary Research Trends in Science, Technology, Engineering, and Mathematics: Studies on Urban Classrooms.*

7. See Lisa R. Lattuca, *Creating Interdisciplinarity: Interdisciplinary Research and Teaching among College and University Faculty* (Nashville, TN: Vanderbilt University Press, 2001); Lisa R. Lattuca, Lois J. Voigt, and Kimberly Q. Fath, "Does Interdisciplinarity Promote Learning? Theoretical Support and Researchable Questions," *Review of Higher Education* 28, no. 1 (2004): 23–48 Project Kaleidoscope, *What Works in Facilitating Interdisciplinary Learning in Science and Mathematics: Summary Report.*

8. Exceptions are made for academic departments at the college that provide a home for multiple disciplines, such as Humanities and Social Science.

9. Many opponents of team-teaching at City Tech take this position because of the stipulation that the workload for co-teaching is based on the percentage of the interdisciplinary course that is integrated. CUNY's John Jay College of Criminal Justice is often used as a supporting example, because their Interdisciplinary Studies Program professors receive the same workload hours for co-teaching as they would for traditional teaching. This is model cannot be currently applied to City Tech because our interdisciplinary courses are not housed in a Department of Interdisciplinary Studies. Our STEM-focused institution course offerings are taught by faculty from all disciplines and are not solely taught by two professors (e.g., Healing the Body: The Visual Culture of Medicine is housed in the humanities department and is taught by professors of art history, nursing, and dental hygiene). We offer different ways to team-teach (i.e., using guest lecturers, trading credits, learning communities, etc.), and an interdisciplinary course is required of all baccalaureate students, which results in the need to offer significantly more interdisciplinary courses.

10. The aforementioned definition and suggestions for reframing a course as interdisciplinary are provided on the homepage of our website, https://openlab.citytech.cuny.edu/ids/

11. To date, this has happened four times; three were members, recommended by administration, who joined after we became a college-wide committee.

12. For a complete list, see Clyde Freeman Herreid, "What Makes a Good Case? Some Basic Rules of Good Storytelling Help Teachers Generate Student Excitement in the Classroom," *Journal of College Science Teaching* 27, no. 3 (1997/1998): 163–165.

13. See Reneta D. Lansiquot, Reginald A. Blake, Janet Liou-Mark, and A. E. Dreyfuss, "Interdisciplinary Problem-Solving to Advance STEM Success for All Students," *Peer Review* 13, no. 3 (2011): 19–22.

14. This pedagogical strategy is described in Reneta D. Lansiquot and Candido Cabo, "Strategies to Integrate Writing in Problem-Solving Courses: Promoting Learning Transfer in an Interdisciplinary Context," in *Proceedings of the 122nd American Society for Engineering Education Annual Conference* (Washington, DC: ASEE, 2015). Also refer to Chap. 4 of the present book.

15. For more information on this program, see Costanza Eggers-Piérola, Bonne August, Cinda P. Scott, Pamela Brown, and Reneta D. Lansiquot, "Promoting an Interdisciplinary Campus Culture," in *Technology, Theory, and Practice in Interdisciplinary STEM Programs: Connecting STEM and Non-STEM Approaches*, ed. Reneta D. Lansiquot (New York: Palgrave, 2016).

16. For a complete list of current courses and sections designated as interdisciplinary, including course descriptions, refer to https://openlab.citytech.cuny.edu/ids/current-courses/

17. The Collaborative on Academic Careers in Higher Education (COACHE) is a Harvard-based consortium of institutional leaders who are taking cost-effective steps to improve outcomes in faculty recruitment, development, and retention. For more information, see http://sites.gse.harvard.edu/coache

18. COACHE Faculty Job Satisfaction Survey, Provost's Report, New York City College of Technology, 2015, http://air.citytech.cuny.edu/air/Surveys/coache.aspx

BIBLIOGRAPHY

Bloom, Benjamin S., Max B. Englehart, Edward J. Furst, Walter H. Hill, and David R. Krathwohl. 1956. "Taxonomy of Educational Objectives, the Classification of Educational Goals." In *Handbook I: Cognitive Domain*, edited by Benjamin S. Bloom. New York: McKay.

Eggers-Piérola, Costanza, Bonne August, Cinda P. Scott, Pamela Brown, and Reneta D. Lansiquot. 2016. "Promoting an Interdisciplinary Campus Culture." In *Technology, Theory, and Practice in Interdisciplinary STEM Programs:*

Connecting STEM and Non-STEM Approaches, edited by Reneta D. Lansiquot. New York: Palgrave.

Elrod, Susan, and Mary J. S. Roth. 2015. "Framing Leadership for Sustainable Interdisciplinary Programs." *Peer Review* 17 (2): 8–12.

———. 2012. *Leadership for Interdisciplinary Learning: A Practical Guide to Mobilizing, Implementing, and Sustaining Campus Efforts*. Washington, DC: AAC&U.

Herreid, Clyde Freeman. 1997/1998. "What Makes a Good Case? Some Basic Rules of Good Storytelling Help Teachers Generate Student Excitement in the Classroom." *Journal of College Science Teaching* 27 (3): 163–165.

Krathwohl, David R. 2002. "A Revision of Bloom's Taxonomy: An Overview." *Theory into Practice* 41 (4): 212–218.

Lansiquot, Reneta D. 2013. *Cases on Interdisciplinary Research Trends in Science, Technology, Engineering, and Mathematics: Studies on Urban Classrooms*. New York: Information Science Reference. doi: 10.4018/978-1-4666-2214-2.

Lansiquot, Reneta D., and Candido Cabo. 2015. "Strategies to Integrate Writing in Problem-Solving Courses: Promoting Learning Transfer in an Interdisciplinary Context." *Proceedings of the 122nd American Society for Engineering Education Annual Conference*. Washington, DC: ASEE.

Lansiquot, Reneta D., Reginald A. Blake, Janet Liou-Mark, and A. E. Dreyfuss. 2011. "Interdisciplinary Problem-Solving to Advance STEM Success for All Students." *Peer Review* 13 (3): 19–22.

Lattuca, Lisa R. 2001. *Creating Interdisciplinarity: Interdisciplinary Research and Teaching among College and University Faculty*. Nashville, TN: Vanderbilt University Press.

Lattuca, Lisa R., Lois J. Voigt, and Kimberly Q. Fath. 2004. "Does Interdisciplinarity Promote Learning? Theoretical Support and Researchable Questions." *Review of Higher Education* 28 (1): 23–48.

Lave, Jean, and Etienne Wenger. 1991. *Situated Learning: Legitimate Peripheral Participation*. Cambridge: Cambridge University Press.

MacDonald, Sean P., Olufemi Sodeinde, and Andleeb Zameer. 2013. "Theories into Practice: A Focus on STEM at City Tech." In *Cases on Interdisciplinary Research Trends in Science, Technology, Engineering, and Mathematics: Studies on Urban Classrooms*, edited by Reneta D. Lansiquot. New York: Information Science Reference.

Piaget, Jean. 1973. *To Understand Is To Invent: The Future of Education*. New York: Grossman.

Project Kaleidoscope. 2011. *What Works in Facilitating Interdisciplinary Learning in Science and Mathematics: Summary Report*. Washington, DC: AAC&U.

Rhoten, Diana, and Stephanie Pfirman. 2007. "Women, Science and Interdisciplinary Ways of Working." *Inside Higher Ed*. Accessed October 22, 2007. https://www.insidehighered.com/views/2007/10/22/rhoten

Spiro, Rand J., Paul J. Feltovich, Michael J. Jacobson, and Richard L. Coulson. 1992. "Cognitive Flexibility, Constructivism, and Hypertext: Random Access Instruction for Advanced Knowledge Acquisition in Ill-Structured Domains." In *Constructivism and the Technology of Instruction: A Conversation*, edited by Thomas M. Duffy and David H. Jonassen, 57–76. Hillsdale, NJ: Lawrence Erlbaum Associates.

Vygotsky, Lev S. 1978. *Mind in Society: The Development of Higher Psychological Processes*. Cambridge, MA: Harvard University Press.

CHAPTER 2

A Study of Integration: The Role of *Sensus Communis* in Integrating Disciplinary Knowledge

Laureen Park

Abstract Integration is an important notion for interdisciplinary studies. Achieving this shows that the interdisciplinary learner has successfully understood the commonalities among disciplines, as well as exercised crucial cognitive skills. This chapter attempts to elucidate how students integrated various disciplinary perspectives in the interdisciplinary course, Weird Science: Interpreting and Redefining Humanity. The study uses Hans-Georg Gadamer's notion of the *sensus communis* to clarify how it was that students were processing and accomplishing the goal of integrating different disciplinary perspectives as evidenced in class observation, discussion, and especially student papers. The study demonstrates the ways in which common sense knowledge conditions and enables the integration process.

Keywords Common sense • Gadamer • Interdisciplinarity • Integration • *Sensus communis* • *Transdisciplinarity*

L. Park (lpark@citytech.cuny.edu ✉)
Department of Social Science, New York City College of Technology,
City University of New York, Brooklyn, NY, USA

© The Editor(s) (if applicable) and The Author(s) 2016
R.D. Lansiquot (ed.), *Interdisciplinary Pedagogy for STEM*,
DOI 10.1057/978-1-137-56745-1_2

Much has been written about interdisciplinarity and its promise for addressing crucial educational objectives. William Newell, an established scholar of interdisciplinary studies, believes that interdisciplinary studies and integrative learning are the two most effective tools for educating students to be prepared for the complex world in which we live.[1] Julie Thompson Klein, another established scholar in interdisciplinary studies, notes that, "The costs of ignoring these commonalities [of the disciplines] are enormous."[2] There are many ways to approach ID and integration, and scholars in the field have attempted to elucidate different aspects of their methods, objectives, function, and history. Not all scholarship in interdisciplinary studies focuses on integration, however. Like Newell, Klein, and other scholars in the field, this study asserts that integration is central to the process and objectives of interdisciplinary studies.[3] It is the successful achievement of this that sets ID apart from multidisciplinarity, an approach that does not necessarily seek commonalities among different disciplines.

This chapter presents reflections on my four years of experience teaching in an introductory level interdisciplinary course offered at New York City College of Technology (City Tech) called Weird Science: Interpreting and Redefining Humanity. The goal of the course is to explore what it means to be human in the age of technology by exposing undergraduate students to various disciplinary perspectives which are provided by guest lecturers. There is one main instructor whose discipline is English, who listens to all the guest lecturers along with the students, and who helps to facilitate the process of integrating the disciplines. The guest lecturers are primarily full-time faculty members at the college, with representation from physics, biology, psychology, computer sciences, sociology, and philosophy, my discipline. Using observations from class lectures and discussions, verbal and e-mail discussions with students, and homework and papers, some conclusions are drawn about how emerging learners integrate disciplinary knowledge and how philosophy in particular supports this process.

Initially, the notable quality of student papers was what was apparently lacking. Papers appeared rather quirky and colloquial. For example, a student, Walter Cadman,[4] wrote, in referring to Descartes' *cogito ergo sum*, "According to Descartes a person that does not have a cognitive process is considered nonexistent."[5] This is, at best, a misunderstanding of Descartes' idea that thinking is proof of existence. Other examples include a few students whose theses claimed that the question of what it means to be human was unanswerable or endless.[6] Several other papers had seemingly subjective and opinionated theses, such as that humans were greedy and selfish or that we should find a new definition of genius.[7]

This initial lens used in interpreting these papers was shaped by an assumption that students would have a ready, academic discourse that could transcend and unify the various disciplines. However, it might be more accurate to say that, despite initial appearances, students were doing their best to integrate disciplinary knowledge, and that they indeed had a ready transdisciplinary discourse available to them to draw from, though this discourse was not primarily academic.

The rest of this chapter argues that students were relying on the *sensus communis* to integrate the insights from the various disciplines, and to construct commonalities about them. This term, borrowed from Hans-Georg Gadamer's *Truth and Method*, emphasizes the technical understanding of 'common sense' to analyze its role in the integration and production of knowledge in the emerging learner of interdisciplinary studies.[8] In the interdisciplinary studies literature, the role of a common sense informed by life and academic experiences is hinted at, but not explicitly treated. As Klein writes, "There is no unique or single pedagogy for integrative interdisciplinary learning... All of these approaches draw from multiple perspectives on a complex phenomenon for insights that can be integrated into a richer, more comprehensive understanding. In integrative learning, perspectives emanate from disciplines, cultures, subcultures, or life experiences."[9] In Weird Science, students were introduced to disciplines that were unfamiliar to them, and relied, perhaps a little too heavily, on their life experiences and communal mores in making sense of them. Presumably, with more time and experience, they will become more versed in academic discourse. In the meantime, the struggles the students faced in this course spoke to some fundamental aspects about the learning process.

Gadamer says that all knowledge is generated out of the concerns and discourse of the *sensus communis*, the collective sense of a community, or sometimes referred to as the ethos of the community or "common knowledge" that is cultivated over time through the shared lived experiences of its members.[10] Gadamer did not originate the concept of *sensus communis* (that honor is usually attributed to Immanuel Kant in philosophical circles), but he helped to merge it with the notion of the *Lebenswelt* (the lived world) that Wilhelm Dilthey, a predecessor of Gadamer, began.[11] Prior to Gadamer's rendering of the concept, it was used in philosophy to account for the universality of the aesthetic sense, or, in other words, the general agreement people have about whether something is or is not beautiful. However, Kant considered the aesthetic sense itself "subjective" and unscientific, and therefore suffered by its comparison to the rational, scientific mind. Elsewhere, the fields of rhetoric, biblical exegesis, and other fields, starting with the eighteenth-century thinker, Giambattista Vico, saw in the concept of *sensus communis* a unique role as a practical, concrete, and active basis for judgment in discourse and interpretation.

In line with the goal of rhetoric, Gadamer revives the notion of *persuasion* as a way of philosophical argumentation that opens up discourse in a way the more precise and narrow *demonstrations* of the natural sciences (*Naturwissenschaften*) does not. For Gadamer, arguments must be convincing to the informed person, which appeals to the broad, universal, and multi-faceted elements of the *sensus communis*. It is this openness and universality that was perceived as vague and merely approximating academic discourse in the student papers in Weird Science. But it is exactly the lived experiences we have in the community that make us aware of and conditions the problems that eventually become a part of the problems of academic discourse. Gadamer calls this precondition of academic knowledge, "foreknowledge."

Below, the problems facing the notion of integration for interdisciplinary learning are examined, via an attempt to more precisely define what integration is, to establish how the individual learner integrates disciplinary knowledge, and to establish the extent and parameters of commensurability and consensus between disciplines. According to J. Britt Holbrook, a neutral field of discourse that transcends disciplinary limitations (i.e., a transdisciplinary discourse) seems to be indicated as the underlying construct for interdisciplinary communication and learning. However, unlike Holbrook, this study argues that the transdisciplinary discourse is based on the *sensus communis*, and not one that is constructed out of a combination of academic disciplines.

THE 'WHAT' AND 'WHO' OF INTEGRATION

Integration is a critical objective in interdisciplinary learning. Accomplishing this goes to demonstrating that the student has found commonalities between disciplines, as well as differences, and shows the exercise of crucial meta-cognitive and critical thinking skills. But what does it mean for disciplines to become integrated? Does it mean that a variety of disciplines become subsumed under one discipline? Does it mean that elements from all the disciplines become blended? Lattuca, Voight, and Fath delineate four ways disciplines can interact in interdisciplinary learning: *informed interdisciplinarity, synthetic interdisciplinarity, transdisciplinarity,* and *conceptual interdisciplinarity.* In informed interdisciplinary studies, a single discipline is the focus, and encounters with other disciplines are used to enhance or elaborate the main discipline of concern. In synthetic interdisciplinary studies, a number of disciplines are considered in searching for commonalities, but the disciplines remain identifiable with their parameters and methods intact. In transdisciplinary, disciplinary learning is not the main concern—one uses an approach that "transcends" the disciplines and provides a neutral, overarching basis in looking at the various disciplines. Finally, in conceptual interdisciplinary studies, the authors conceive of an interdisciplinary idea that seeks to explore and comprehend fundamental concepts of experience and their limits. This approach might use disciplinary learning as a tool, but is not focused on a single discipline.[12]

The Weird Science course exhibited characteristics of the latter three ways of interacting. There was definitely an expectation that students would understand disciplinary methods and vocabulary on their own grounds, but there was also an expectation that students would be able to unify them in an overarching way in their papers. Students rely on the *sensus communis* as a transdisciplinary, extra-disciplinary approach to interdisciplinary learning, as well as the basis for understanding the fundamental concepts of disciplines. The important question in terms of integration is on what basis, whether methodologically or conceptually, do students unify these perspectives? Moreover, how does the individual learner go about integrating the different perspectives?

The process that goes on at the individual learner's level is an important area to focus on in a discussion of integration. However, the literature is surprisingly vague on the topic of the 'who' of integration. Two philosophers, Hanne Andersen and Susann Wagenknecht, explore this question: "Who is ... the knowing subject of interdisciplinary teamwork in the end?"[13] What is their concluding thought on the answer? It is this: "There is no general answer to this question."[14]

The reason for the difficulty in answering the question of who is the subject of interdisciplinary learning is probably obvious. In interdisciplinary learning, there are layers of complexity involving both the individual learner, as well as the various players involved in the process—the experts and the group members and their roles in the integration process. The authors, citing research done by F.A. Rossini and A.L. Porter, isolate four primary ways experts and group members interact in interdisciplinary research: "integration by leader," "common group learning," "negotiation among experts," and "modeling."[15] It is not worth delving into the intricacies involved in each category of interactions since they mainly involve interactions among experts toward research, which is not the main concern of this chapter. The main question here, whether the interdisciplinary work is based on modeling, or on an expert leader or the group members themselves, is how the work achieves what the authors call "cognitive integration" in the individual researcher or learner, "[N]amely that it will often be too demanding for the individual group member to master several fields in detail and that it will therefore often lead to a decrease in depth."[16] In the end, the authors have trouble establishing how a learner integrates interdisciplinary knowledge on an individual level, and leaves it an open question as reflected in the above quote. But is this not the crucial issue to be resolved when it comes to the issue of the integration of interdisciplinary knowledge?

CONSENSUS, COMMENSURABILITY, AND THE NEED FOR A TRANSDISCIPLINARY DISCOURSE

In the interdisciplinary literature, despite the importance of the notion of consensus, it is often presumed rather than treated explicitly. But, as Holbrook shows well, we cannot presume that discourse, and therefore understanding, between disciplines is even commensurable, much less unifiable under a universal discourse. Clearly, discourse between disciplines is not always smooth and transparent. A biologist does not necessarily understand the nuances of quantum mechanics, nor does a physicist understand the nuances of genetic theory.

Holbrook identifies the "Habermas-Klein thesis" as the best explanation of the integration between disciplines that dominates interdisciplinary scholarship. It holds that the commensurability and consensus between disciplines can be achieved in a way that is transparent to all involved. In this thesis, the desirability and possibility of consensus is mostly presumed, and the transparency of communication and reason is mostly unproblematic. This thesis draws its inspiration from Jurgen Habermas' model of communicative action in which rational actors, based on trust, rational utterances, and reciprocation come to a common understanding. Any obstacles to a common, rational understanding are seen as temporary and commensurable. Holbrook believes that this take on communication describes Klein's view of interdisciplinary communication as well.[17] While Klein, a well-known figure in interdisciplinary scholarship, acknowledges the nuances and complexities of interdisciplinary learning, she nonetheless remains committed to the view that a common understanding based on this commensurability can be achieved. Holbrook believes that actual interdisciplinary communication is messier and more complex than the picture depicted by the Habermas-Klein thesis.

In comparison, Holbrook's Kuhn-MacIntyre thesis holds that disciplines evolve into language systems with their own paradigms and conceptual frameworks, and is incommensurable to each other. This thesis draws its inspiration from Thomas Kuhn, who famously argued that scientific paradigms were a historical and conventional creation. Once created, he argued, a scientific paradigm (e.g., Einstein's relativity model of the universe) began to define the meaning and significance of the language a scientist used. For Kuhn, the conventional and unscientific paradigm preceded and set the stage for scientific work. The implication of the Kuhn-MacIntyre thesis for interdisciplinary studies is that disciplinary communication cannot be truly integrated, and that instead they can merely interact with each other. According to this thesis, no one working in one discipline could truly understand the meanings and significance of the language of another discipline without being inculcated in that discipline's worldview. Learning the language of a discipline is like learning a foreign language. In this case, integration is seen as possible only as translation of one disciplinary discourse into another. This can be seen as a case of synthetic interdisciplinary studies in the way Lattuca, Voight, and Fath spoke of. The situation does not support the notion that a neutral approach could guide the discourse, nor that a blended one could be achieved. The disciplines can be interdisciplinary by co-existing and being in dialogue with each other, even while maintaining each its own parameters and methods.

Holbrook's Bataille-Lyotard thesis is most useful for the present argument. In this thesis, Holbrook makes a distinction between "weak" and "strong" communication, appropriating and reinventing George Bataille's usages of the terms. He calls communication "weak" when everyone involved in the communication readily understands the language and there are no strong barriers blocking communication. It is "weak" because the language is penetrable. Holbrook understands this form of communication as the predominant one in communication within disciplines, but also implies that the same commensurability holds, for the most part, in interdisciplinary communication.[18] The present argument shows that the reason commensurability holds between disciplinary discourse is because of a more fundamental commonality owing to the *sensus communis*. Indeed, Holbrook himself seems to indicate that the goals of weak communication is based on the concerns of the community at heart: "Bataille's weak communication is thus used for the purposes of gaining a clear understanding of the things that constitute the objective world and of establishing a consensus as to how we ought to act in order to be productive members of society."[19] Reflecting this idea, Gadamer shows below that our concerns and problems as members of a society and as people defined by the human condition are at the heart of what we study in academic disciplines. However, once the problems become removed from the context of lived experience and become a part of the abstract framework of disciplinary discourse, common sense starts to appear inadequate to elucidate those same problems.

The other form of communication that Holbrook attributes to the Bataille-MacIntyre thesis is the "strong" form of communication. This is when language presents barriers to communication between disciplines. In Bataille's formulation, weak communication cannot penetrate strong communication when the discourse begins to express extraordinary concerns that transcend our ordinary understanding and therefore causes a break in ordinary communication. To understand Bataille's understanding of strong communication adequately would take us too far afield. The important thing here is to understand Holbrook's appropriation of this term. In Holbrook's context, an example of strong communication might be over-specialized disciplinary terminology. A learner outside the discipline might follow specialized discourse up to a point, but communication would break down if it became so specialized that the learner's conceptual framework was no longer adequate to making sense of the discourse. Nonetheless, the productive aspect of strong communication is that it highlights areas of difference between disciplines, which is as important a characteristic in interdisciplinary studies as is the commonalities underlying the disciplines. Holbrook sees this thesis' accounting of difference as an improvement over what he believes to be the naïveté of the Habermas-Klein thesis.

Holbrook's solution to the problem presented by strong communication reflects this study's view about how communication works, although not completely. He believes that the solution to a breakdown in interdisciplinary communication is to "co-create" a new genre of discourse that discards the identities of the old disciplines.[20] Essentially, we need a transdisciplinary approach to interdisciplinary discourse. However, what would a genre of discourse look like that both blends and blurs the identities of several disciplines? Holbrook does not fully elaborate what this would entail. Holbrook's goal might make more sense applied to more specialized academic work. Here, Gadamer's notion of the *sensus communis* seems to better elucidate the transdisciplinary approach needed to understand how students in an interdisciplinary course integrated disciplinary discourse.

THE ROLE OF THE *SENSUS COMMUNIS* IN INTERDISCIPLINARY INTEGRATION

Gadamer believed that our training as learners begins with our immersion in the communal sense of the *sensus communis*. This communal sense is not interested in precision or specialization, but dwells in the "verisimilar" or probable.[21] To the specialist, this imprecision might be seen as vague and inadequate. But for Gadamer, it speaks to the fact that common sense is broad and open enough to be the flexible source of the breadth and depth of the full scope of academic disciplines. Whether we become specialists in the stars above or in the psychological dynamics within, our first approximations about how the world works all start in common sense.

For Gadamer, the most basic level of educating someone is in acculturating that person into the *sensus communis* "in getting beyond his naturalness."[22] We are born biological creatures, but we *become*, through experience, acculturated human beings with ethical, aesthetic, intellectual concerns that go beyond the merely natural. From the earliest associations with others, we are always and already in the midst of being acculturated into the habits, norms, and traditions of our community. He calls this process the process of *Bildung* (culture or formation).[23] In the earliest forms, the knowledge we get from *Bildung* is approximate and more unconscious than reflective, but, as we become more educated within more narrowly defined "cultures" of academic disciplines or other arenas of learning, the knowledge becomes more precise, reflective, and/or rational. Though the quality and the quantity of the knowledge may change, the fundamental process of acquiring education, for Gadamer, remains the same throughout. He writes, "Hence, all theoretical *Bildung*, even acquiring foreign languages and conceptual worlds, is merely the continuation of a process of *Bildung* that begins much earlier."[24]

As noted, the process begins with the *sensus communis*, the communal sense that we all share in our lived experiences with others. There are two aspects of the *sensus communis* that are significant for our purposes in looking at how students integrate disciplinary knowledge. First, the *sensus communis* is inherently consensus-building, and second, it has an inherent sense of standards that seek to evaluate and validate knowledge. These two aspects of the *sensus communis* condition our higher-order thinking in the disciplines.

A word here about what it means to share a communal sense would be useful. Not all communities share the same norms and expectations, and, clearly, there is a wide variation of what is acceptable, especially from a global perspective. What is considered culturally normal in Korea will, of course, differ from what is considered culturally normal in Turkey. Even in one country, like America, there can be subcultures within a dominant mainstream culture, as well as fluidity between subcultures. Gadamer faced criticism about what he meant by culture and tradition, criticisms that he could not fully address. It is doubtful that anyone who speaks about culture could fully address what does and does not count toward inclusion in a given culture. To do so, in any case, would be beyond the scope of this chapter.

Instead, Gadamer focuses on the ways in which our experiences in the community contribute to the font of knowledge, which in turn provides the basis for our larger, more speculative ideas about how the world works and what our place in it is. In other words, he is interested in the shared collection of knowledge that is applicable toward understanding the human condition.[25] The main point here is that Gadamer's concerns are quite distinct from those of an anthropologist or someone working in another discipline. He is not so interested in the empirical or "natural" concerns of members of a community—how they acquire food, how they heal sick people, or what their marriage and burial rituals are. In alignment with Gadamer's focus, student papers in this interdisciplinary course did not focus so much on the particularities of their lives as on broad, sweeping concerns, indicating that students were interested in exploring the deep, fundamental issues of human experience.

When they had one, students' theses clearly reflected concerns that are in line with a thoughtful person's reflections on the human condition. A few students expressed hope that the knowledge gained in the course would be learned by all for the sake of a better future for humanity.[26] Other students expressed concerns that humans are incorrigibly greedy or power-hungry.[27] These were themes borne out of their lived experiences and which were familiar to them. From these frameworks, they could then construct a bridge to the more unfamiliar discourse of the academic experts. Their attempts at transforming pre-existing knowledge to the new ones had mixed success, but their attempts were as much as could be expected from introductory students.

For Gadamer, the work of the academic is the culmination and fulfillment of the same impulses that motivate common sense understanding. But whereas the valuations of the person of common sense only assert itself "without being able to give its reasons,"[28] scholars are able to be self-reflective about their assumptions and assertions. The fact that disciplines can become so specialized speaks to the level of sophistication and nuance that people are capable of generating and advancing. This is the mark of human ingenuity for Gadamer, and need not be seen as antagonistic to common sense or elitist as some philosophers believe.[29]

Gadamer describes the *sensus communis* as both historical and aesthetic. What he means is that communities share habits, customs, and traditions that are connected by a shared history. This history informs the worldview (*Weltanschauung*) of the community's members and shapes our values and the very way we see our world. This accumulated/constructed history at any given time makes up our culture. In addition to it being historical, Gadamer also describes the *sensus communis* as "aesthetic" because, initially, our sense of what is right, wrong, good, and beautiful is based on an affective sense governed by the norms and customs of our community, which are not necessarily rational, academic, or scientific. Only after more experiences with formal learning does one begin to account for and explain their valuations. Even then, Gadamer nonetheless maintains that the various scholarly disciplines within which scholars work are no less communities with habits and traditions as is the *sensus communis*. This justifies why Gadamer later shifts the talk of aesthetics to *prejudice* in developing his method for understanding the human sciences.[30] For Gadamer, like common sense, disciplines are self-justifying and rely on a history of created norms or traditions, or prejudices. Disciplines, however, have less membership than the common sense community, and are, in this way, more specialized. The task for the emerging learner of interdisciplinary studies is to integrate the more specialized language of disciplines into the more common discourse, and, in turn, become acculturated into the new discourse. Fortunately, as Gadamer has been arguing, there is much foreknowledge that is already shared between common and theoretical discourses.

Two fundamental features of the *sensus communis* that carry over into theoretical discourse can also help us to understand what we have said about integration. First, Gadamer accounts for the commensurability of discourse, as well as the unifying element of integration by revealing a circularity in the process of acculturating individuals in the *sensus communis*, which is the same process in acquiring disciplinary knowledge. The individual uses what is already familiar to condition the reception of new knowledge, and seeks, in turn, to transform the new knowledge into the familiar. Gadamer writes, "To recognize one's own in the alien, to become at home in it, is the basic movement of spirit [*Geist*], whose being consists only in returning to itself from what is other."[31]

Second, common sense seeks to make sense of the world; in the most minimal expression of this making sense, a standard is implied. Prior to mature judgment, Gadamer believes that we evaluate the world in terms of *tact* or *taste*. "Tact" is often used to refer to behaviors and "taste" to an aesthetic judgment, but both are acquired senses that make judgments using standards learned from one's community. A child might say that he should not speak too loudly because "mother says so," or that he likes the look of those shoes "just because." They are "modes of knowledge," Gadamer says, but ones that rely on a standard that is merely felt. With maturity, experience, and more education, one can make more reflective judgments, using a stronger sense of *validity*.[32] Nonetheless, that standard, for Gadamer should remain open and flexible, reflecting the human experience.[33]

These two observations about the relationship between common sense and disciplinary knowledge help to elucidate how students in the interdisciplinary course in fact integrated the various disciplinary perspectives presented to them. No matter what the discipline, there was an utter conviction that the disciplines had something to offer the students, and that they would understand the significance of the lessons. They also approached the disciplines with the conviction that they were capable of evaluating the worth of the lessons, rooted in the expectations that their experience in the common sense world provided.

INTEGRATION IN STUDENT TERM PAPERS

The most integrated papers in Weird Science also exhibited the most dissonance. That might not be such a bad thing. According to John C. Bean, who wrote a popular guide book for professors to encourage active learning in the classroom, "cognitive dissonance" should be the very objective of our pedagogical methods.[34] For, in this dissonance, students' familiar ways of thinking are challenged, and thus awakened, become open to new insights. The more ambitious a student was in integrating disciplinary insights, the more he or she appeared unorthodox and perhaps disorderly. Interestingly, this is in line with Steve Fuller's idea about "deviant" interdisciplinary studies. He argues that, when a thinker attempts to approach interdisciplinary studies outside of the specialized language of academia, she appears "'eclectic' and 'arbitrary', very much as upstart entrepreneurs look to managers in established firms, where the former wish to 'creatively destroy' and the latter to 'monopolize' markets."[35] In other words, academic training sets up an expectation about how students will resolve problems. Using common sense discourse will, by comparison, appear arbitrary and quirky. Inevitably, though, students in the Weird Science course approached problems in the course with the only tools they had, and this was their common sense informed by their life and academic experiences. Their efforts—because of their eclectic approaches and not despite them—suggested to me that they were genuinely engaged in what the various disciplines offered. On the other hand, papers that stuck with one discipline or one theme without attempting to integrate the many disciplinary perspectives offered in the course were also the smoothest and most organized.

This reflection concludes with thoughts on papers from each category: papers attempting to integrate disciplinary perspectives, and papers not attempting to integrate disciplinary perspectives. In general, integrated papers were those wherein the student attempted to provide a unifying theme or thesis that purported to organize ideas from several disciplines. Unintegrated papers were those wherein the student did not attempt to make sense of several disciplinary perspectives.

In terms of integrated papers, Walter Cadman's paper (a paper referred to earlier as misappropriating Descartes) exhibits characteristics of an emerging learner of interdisciplinary studies in integrating new disciplinary insights. His paper, as well as many of the integrated papers, in fact, touches on deep, fundamental matters. Cadman's thesis was about the unique and dominant status that humans had on earth. In discussing his thesis, he points to evolutionary theory, genetic theory, Cartesian philosophy, music, the epic of Gilgamesh, religion, gender theory, prosthetics, physics, and much more. His paper is eleven pages long, which is only space enough for broad, surface renderings of these topics. Nonetheless, it helps him to make a number of conclusions about his thesis, which reflect Gadamer's ideas about the kind of agency the learner has—the agent is the person of *phronesis*, and has both an open-ended and practical comportment to his world. Cadman concludes that "humans will endlessly expand their knowledge"[36] and that "being human is a natural ability that any person should hold dear to themselves and use what is given to express try [*sic*] to better the rest of the living world."[37] A number of students reflected both the "endlessness" of knowledge, as well as expressed ethical aspirations in their conclusions.[38]

In terms of papers that do not attempt to integrate the various disciplinary perspectives, Gregory Martin's paper was quite cohesive and well-argued, two marks of a well-written paper. But it also did not, by the same token, take much risk in terms of exploring perspectives that were not new to him. Despite the fact that Martin made a conclusion that was reminiscent of a "deviant" interdisciplinary paper, writing, "[T]he question should not be, 'What does it mean to be human?' the question should be, 'What could humans do to bring peace amongst ourselves?' would the wealthy human's response be the same of a homeless humans' response? [*sic*],"[39] the actual content of the paper revolved around the theme of the human brain and the scientific understanding of it. Martin begins with a discussion about the genetic similarity between humans and chimpanzees, and then goes on to touch upon how a doctor used HeLa cells therapeutically, how we evolved from the Peking man, how we have changed the physical environment of the earth, and how we can make artificial meat from 3-D printers, in addition to other scientific topics related to the brain. Martin does draw from a variety of topics, but they mostly relate to the biological sciences. Despite the sweeping question that ends his paper, Gregory Martin's paper presents a narrow argument which does not contain much that sticks out as particularly arbitrary or quirky.

Conclusion

This chapter started with the premise that integration was a key notion for interdisciplinary studies, elucidating how integration happened in the emerging learner of interdisciplinary studies as evidenced by my interaction with students in the Weird Science course. The key figure that elucidated this process was Gadamer and his ideas about the *sensus communis* wherein all disciplinary knowledge is the outgrowth of a more fundamental acculturation process that begins with common sense knowledge.

Two significant aspects of common sense reveal the conditions of how we learn anything at all. First, it seeks to unify new knowledge into the familiar store of common sense knowledge, which explains the unifying and consensus-seeking aspect of integration. Second, it is inherently governed by a standard of validity, which speaks to the fact that one integrates knowledge based on the norms and values of one's community. Later, the standards of validity become heightened as the community becomes more specialized and disciplinary.

Another theme of this chapter was to validate the integration process that the Weird Science students exhibited in their papers. Initially, it was difficult to understand how it was that students were integrating the various disciplinary perspectives in their pursuit of the question, "What does it mean to be human?" Through the interpretation that they were using common sense (per Gadamer) as a basis of that integration, their unorthodox and eclectic ways of interpreting the problem in their papers was a sign of something more productive of the learning process. It was a sign that they were genuinely engaging with what was unfamiliar to them, which caused them to appropriate the new ideas in ways that seemed quirky from the standpoint of a disciplinarian. In contrast, those who chose to use more finite, perhaps familiar, disciplinary parameters produced smoother, more organized papers (i.e., among those who attempted to write a thesis-driven, organized paper). Presumably, with time and experience, all the students will go on to become more versed in academic and disciplinary language, just as all professional academics have already done.

NOTES

1. William Newell, "Educating for a Complex World: Integrative Learning and Interdisciplinary Studies," *Liberal Education* 96, no. 4 (Fall 2010): 6.
2. Julie Thompson Klein, *Interdisciplinarity: History, Theory, and Practice* (Detroit: Wayne State University Press, 1991), 14.
3. Newell, "Educating for a Complex World," and Klein, *Interdisciplinarity.*
4. All student names noted in this chapter are pseudonyms.
5. Walter Cadman, "On Being Human," *Weird Science* paper (New York: City Tech, 2015), 5.
6. Cadman, "On Being Human", Jennifer Mendez, "What Does It Mean to be Human?" *Weird Science* paper (New York: City Tech, 2015); Leon Perez "Motivation," *Weird Science* paper (New York: City Tech, 2015).

7. Diane King, "Future Genius Generation," *Weird Science* paper (New York: City Tech, 2015); Richard Smith "Misery Enjoys Selfishness," *Weird Science* paper (New York: City Tech, 2015).

8. Hans-Georg Gadamer, *Truth and Method*, trans. Weinsheimer, Joel and Marshall, Donald G. (New York: Continuum, 1995).

9. Julie Thompson Klein, "Integrative Learning and Interdisciplinary Studies," *Journal of the AAC&U* 7, no. 4 (2005): 9.

10. Klaus Dockhorn, "Hans-Georg Gadamer's 'Truth and Method," *Philosophy & Rhetoric*, trans. Marvin Brown, 13, no. 3 (1980): 160–180.

11. Jeff Polet, "Taking the Old Gods with Us: Gadamer and the Role of Verstehen in the Human Sciences," *The Social Science Journal* 31, no. 2 (1994): 171–196.

12. Lisa R. Lattuca, Lois J. Voigt, and Kimberly Q. Fath, "Does Interdisciplinarity Promote Learning? Theoretical and Researchable Questions," *The Review of Higher Education* 28, no. 1 (2004): 25–26.

13. Hanne Andersen and Susann Wagenknecht, "Epistemic Dependence in Interdisciplinary Groups," *Synthese* 190, no. 11 (2013): 1896.

14. Ibid., 1896.

15. Ibid., 1888.

16. Ibid., 1890.

17. Holbrook, "What is Interdisciplinary Communication?" 1869–1870.

18. Ibid., 1877.

19. Ibid., 1874.

20. Ibid., 1878.

21. Gadamer, *Truth and Method*, 20–21.

22. Ibid., 14.

23. The usual English translation for "*Bildung*" is "formation" or "education," and "*Kultur*" is usually interpreted as "culture." However, the convention followed in Weinsheimer's translation of *Truth and Method* is used here. For Gadamer, becoming educated is not just about learning as we conventionally think of it, but an ontological process of becoming human in the ethical, practical, and intellectual sense. The collective formative process is culture and the individual is acculturated therein.

24. Ibid.

25. In philosophy, there is a certain amount of discomfort in seeking universal commonalities across cultures, for such attempts have been criticized for silencing diversity, especially perspectives from non-European points of view. In addressing such criticisms, Gadamer has conceded that his approach might not elucidate traditions outside his own Eurocentric tradition. Either way, if Gadamer is right that his argument rests on persuading the informed person, then readers can decide whether Gadamer's ideas apply to their own experience.

26. Cadman, "On Being Human"; King, "Future Genius Generation."
27. Yasmin Abramovich, "Term Paper," *Weird Science* paper, (New York: City Tech, 2015); Richard Smith, "Misery Enjoys Selfishness"; Duane Phillips, "Abuse of Power: Failures of Authority Leadership and the Profit Motive in Journalism," *Weird Science* paper, (New York: City Tech, 2015).
28. Gadamer, *Truth and Method*, 17.
29. For a fascinating critique of the specialization of disciplinary language as a detriment to exploring the human condition in a broader, more fundamental way, see Steve Fuller, "Deviant Interdisciplinarity as Philosophical Practice: Prolegomena to Deep Intellectual History," *Synthese* 190, no. 11 (2013): 1899–1916.
30. As in Gadamer, *Truth and Method*, 271.
31. Ibid., 14.
32. Ibid., 36.
33. Ibid., 17.
34. John C. Bean, *Engaging Ideas* (San Francisco: Jossey-Bass, 2001), 27.
35. Fuller, "Deviant Interdisciplinarity," 1902.
36. Cadman, "On Being Human," 10.
37. Ibid., 11.
38. E.g., Leon Perez, "Motivation"; Menendez, "What Does It Mean to be Human?"; Stephen Devis, "Final Report," *Weird Science* paper (New York: City Tech, 2015).
39. Gregory Martin, "What Does It Mean to be Human? Is that the Question?" 9.

Bibliography

Andersen, Hanne, and Wagenknecht, Susann. 2013. "Epistemic Dependence in Interdisciplinary Groups." *Synthese* 190 (11): 1881–1898.
Bean, John C. 2001. *Engaging Ideas*. San Francisco, CA: Jossey-Bass.
Dockhorn, Klaus. 1980. "Hans-Georg Gadamer's 'Truth and Method." Translated by Marvin Brown. *Philosophy & Rhetoric* 13 (3): 160–180.
Ferrara, Alessandro. 2008. "Does Kant Share Sancho's Dream?: Judgment and *Sensus Communis.*" *Philosophy Social Criticism* 34 (1–2): 65–81.
Fuller, Steve. 2013. "Deviant Interdisciplinarity as Philosophical Practice: Prolegomena to Deep Intellectual History." *Synthese* 190 (11): 1899–1916.
Gadamer, Hans-Georg. 1995. *Truth and Method*. Translated by Joel Weinsheimer and Donald G. Marshall. New York: Continuum.
———. 1980. *Dialogue and Dialectic: Eight Hermeneutical Studies on Plato*. Translated and edited by P. Christopher Smith. New Haven, CT: Yale University Press.

———. 1976. *Philosophical Hermeneutics.* Translated and edited by David Linge. Berkeley: University of California Press.

Gueorguieva, Valentina. 2003. "Les deux faces du sens commun." *The Canadian Review of Sociology and Anthropology* 40 (3): 249–256.

Holbrook, J. Britt. 2013. "What is Interdisciplinary Communication? Reflections on the Very Idea of Disciplinary Integration." *Synthese* 190: 1865–1879.

Klein, Julie Thompson. 1991. *Interdisciplinarity: History, Theory, and Practice.* Detroit, MI: Wayne State University Press.

———. 2005. "Integrative Learning and Interdisciplinary Studies." *Journal of the AAC&U* 7 (4): 8–10.

Lattuca, Lisa R., Lois J. Voigt, and Kimberly Q. Fath. 2004. "Does Interdisciplinarity Promote Learning? Theoretical and Researchable Questions." *The Review of Higher Education* 28 (1): 23–48.

Newell, William. 2010. "Educating for a Complex World: Integrative Learning and Interdisciplinary Studies." *Liberal Education* 96 (4): 6–11.

Nikitina, Svetlana. 2006. "Three Strategies for Interdisciplinary Teaching: Contextualizing, Conceptualizing, and Problem-Centring." *Journal of Curriculum Studies* 38 (3): 251–271.

O'Rourke, Michael, and Crowley, Stephen. 2013. "Philosophical Intervention and Cross-Disciplinary Science: The Story of the Toolbox Project." *Synthese* 190: 1937–1954.

Polet, Jeff. 1994. "Taking the Old Gods with Us: Gadamer and the Role of Verstehen in the Human Sciences." *The Social Science Journal* 31 (2): 171–196.

Repko, Allen F. 2008. *Interdisciplinary Research: Process and Theory.* Thousand Oaks, CA: SAGE.

Schaeffer, John D. 1981. "Vico's Rhetorical Model of the Mind: '*Sensus Communis*' in the '*De nostri temporis studiorum ratione*'" *Philosophy & Rhetoric* 14 (3): 152–167.

Slavich, George, and Zimbardo, Philip. 2012. "Basic Principles of Transformational Teaching" *Educational Psychological Review* 24 (4): 569–608.

CHAPTER 3

Insatiability and Crisis: Using Interdisciplinarity to Understand (And Denaturalize) Contemporary Humans

Sean P. MacDonald and Costas Panayotakis

Abstract Collaboration between different social sciences can encourage students to think critically about prevailing assumptions regarding human nature. Both the chapter and the pedagogical experiences on which it is based investigate the distinctive type of human created by capitalist society. In so doing, it takes a heterodox approach to analyzing the concept of an insatiable human nature through a case study that invites students to critically assess this perspective. This discussion then leads to an investigation and critique of traditional neoclassical Economic assumptions about human behavior, which forms the basis for a case study on the causes of the global economic and financial crisis of 2008. The goal is to facilitate students' development of a more grounded perspective on real world events.

Keywords Human behavior • Human nature • Neoclassical assumptions • Neoclassical economics

S.P. MacDonald (smacdonald@citytech.cuny.edu ✉) • C. Panayotakis
Department of Social Science, New York City College of Technology,
City University of New York, Brooklyn, NY, USA

© The Editor(s) (if applicable) and The Author(s) 2016
R.D. Lansiquot (ed.), *Interdisciplinary Pedagogy for STEM*,
DOI 10.1057/978-1-137-56745-1_3

37

This chapter explores how an interdisciplinary pedagogical approach can effectively challenge accepted systems of beliefs and pose alternative perspectives that encourage students to think critically about prevailing assumptions regarding human nature. It also illustrates how collaboration between different social sciences—in this instance, Sociology and Economics—can inspire students to investigate and question the distinctive type of human as shaped by capitalist society. Starting from an overview of the beliefs about human nature and behavior as postulated in neoclassical economic theory, we then begin to explore with students how these theoretical constructs naturalize patterns of human behavior that are historically and socially conditioned.

Neoclassical economics defines rational human behavior as characterized by insatiable wants and desires and an attempt to attain efficient outcomes in the face of scarce resources. Economic "agents"—consumers, businesses, and government—are utility maximizing and thus seek to maximize their own self-interest. The choices that emerge from these motivations are said to be "rational." Further, unimpeded competition results in the most efficient distribution of scarce resources. All economic agents are presumed to have all relevant information necessary to guide them in making perfectly informed rational choices. These assumptions are central to neoclassical economic theory and have been applied to the study of consumer/household behavior, competitive business practices, and government decisions about how to allocate resources (i.e., funds) to competing social and economic needs.

These same assumptions are pervasive in standard introductory Economics texts and teachings. In fact, the supposition that "economic agents" act as rational decision makers is built into the way economic participants are expected to behave in consumer-centered economies such as our own. These assumptions can be traced back to the theoretical works of William Stanley Jevons, Alfred Marshall, and Leon Walras in the late nineteenth and early twentieth centuries.[1] More contemporary economists such as Alan Friedman and Robert Michael and Gary Becker[2] have shaped their analyses of human behavior, drawing extensively from these early neoclassical economists. Even with the relatively more contemporary incorporation of Keynesian[3] (primarily) and other economic perspectives and analyses in the study of problems such as unemployment and challenges to macroeconomic growth, as well as monetary and fiscal policy, neoclassical choice theory is still largely grounded in the assumption that policy choices are rational and decisions are constrained by scarce resources. That "scarcity" itself may—at least in part—be created by past policy decisions is rarely addressed.

The institutionalized acceptance of this mainstream theory of human behavior is then said to inform choices and actions in business and financial markets in the quest to arrive at "efficient" outcomes. Finally, decision makers are assumed to operate in the context of market conditions that can supposedly be precisely anticipated and known, much like a laboratory experiment in which all variables can be controlled. Because of this, decision makers can also be assumed to be making perfectly informed decisions with a predictable outcome.

WHAT IS THE MEANING OF THE CENTRAL ASSUMPTIONS UNDERLYING NEOCLASSICAL ECONOMIC THEORY?

How is rational behavior and decision making understood? Utility-maximizing consumers are said to make rational decisions when those decisions are informed by all available information—ideally complete information. For instance, consumers seek to pay the lowest possible price for comparable products, everything else constant (preferences, income, prices of substitutes). Accordingly, they weigh the costs and benefits of a given option, and choose based upon whether the benefits exceed the costs, thus producing the expected best outcome. Such a method of decision making is considered rational and efficient in that the individual is making choices that maximize satisfaction and minimize costs.

The attainment of efficient use of scarce resources is assumed to be the goal of decision making by all economic agents. For the business, efficiency requires the choice of a production methodology that combines resources—labor, capital, and natural resources—in a way that results in the lowest marginal costs (or costs per unit of output) and maximizes marginal revenue or earnings. As such, the efficient choice is one that yields the greatest output from available resources at the lowest possible cost in the idealized model of competitive capitalism. By extension, the choice that emerges from weighing the costs and benefits of various options and choosing the methodology that minimizes costs while maximizing gain is rational.

The existence of insatiable wants and desires forms the foundation of the concept of scarcity and is treated as part of the natural human condition. Not only is it impossible to satisfy one's wants and desires because of the inherent human need to consume, but the presumption of scarcity as a fundamental given makes the attainment of such wants impossible. Thus, consumerism is viewed as a characteristic that both defines human nature and can never be fully satisfied because of the natural existence of insufficient resources. Economies that rely upon the spending of consumers as the central engine of economic growth depend upon a steady flow of income and wealth from consumers to businesses. By appealing to the exclusivity of "desirable" consumer goods, marketing campaigns have effectively cultivated the desired wants and desires of consumers.

While a naturally occurring lack of critical resources such as water in arid climates or soil suitable for growing crops is a reality in some geographic areas, the concept is often conceived of as a universal given. The notion that scarcity of natural resources may result from past decisions about how resources are allocated or used (or misused) rarely surfaces. For instance, public policy decisions may have deliberately created a scarcity of funds for critical human needs, achieving the redistribution of income and wealth toward the wealthy and away from the poor and middle class.

Finally, a fundamental conviction at the heart of the neoclassical theory of competitive capitalism is the belief that unimpeded competition results in the most efficient distribution of scarce resources. In the ideal world, the government has a limited regulatory role in industry and financial markets, which stems from the premise that markets naturally find their equilibrium position. According to this model, if unemployment is too high, wages will fall and employers will hire once again, as increasing supply stimulates and restores growth in demand. In its more contemporary form, this set of assumptions can be linked with the revival of "supply side" economics during the early 1980s.[4]

This critical discussion of these assumptions is important for the purpose of the course's theme, not just because it introduces students to the assumptions that predominate in one of the social scientific disciplines seeking to shed light on the human condition. In contemporary capitalist societies, neoclassical assumptions are often received as the obvious, "common sense" way to understand economic life. This is both because this way of analyzing economic life dominates mainstream media and because of the relative lack of pluralism within the discipline of Economics. This lack of pluralism is especially felt in introductory college-level courses, which usually do not present neoclassical economics as one of the possible ways of analyzing economic life, but rather as *the* economic approach to the analysis of human life. Thus, while a student taking an introductory class in another discipline, for example sociology, would be exposed to various theoretical perspectives, ranging from more conservative functionalist approaches to more progressive or even radical approaches, such as conflict theory, Marxism, and feminism, a student taking Economics 1101 (Introductory Macroeconomics) will usually have no way of knowing that the "Introduction to Economics" course s/he thinks s/he is taking is really an "Introduction to *Neoclassical* Economics."

Given the importance of economic forces in shaping human life and human beings themselves, this is a problem. Students cannot reach a critical understanding of what it means to be human without a critical understanding of economic life. There is also something paradoxical in the lack of pluralism within Economics and especially the lack of pluralism in the way Economics is usually taught to laypeople. On the one hand, neoclassical economics valorizes choice and attributes the alleged superiority of competitive capitalism to the ability it gives consumers to choose between competing versions of the same commodity. On the other hand, neoclassical economists enforce an effective monopoly when it comes to their line of business, the teaching of Economics. And the result of this monopoly is as disastrous as the results of the monopolies that neoclassical economists routinely lambaste. Indeed, the claim that monopoly reduces the pressure to provide top quality products is no less true for the economics profession than it is for other industries. One need only look at the recent financial crisis, which caught people off guard, precisely because their sense of how the economy works came from the hegemonic neoclassical approach which has long taught that nothing can go wrong as long as markets are free.

While presenting to the students the concept of *homo economicus* as postulated by neoclassical economics, we then encourage students to think critically about this model of humanity through a two-stage process. First, one of the authors of this chapter, Costas Panayotakis, encourages students to probe the human insatiability assumption through a discussion that historicizes human needs, while the other author, Sean P. MacDonald, proceeds by encouraging students to evaluate the neoclassical "rationality" assumption in light of the dynamics that led to the global financial crisis in 2008.

CASE STUDY I: QUESTIONING HUMAN INSATIABILITY

Before the session described in this chapter, the students are assigned readings and videos while also being asked to answer questions that deal with both the social construction of human needs and the causes of the recent financial crisis. The readings for the session's discussion of human insatiability include a Reuters article, "U.S. Millionaires Say $7 Million Doesn't Make You Rich, Survey Says"; "The Original Affluent Society," anthropologist Marshall Sahlins' classic essay on hunter and gatherers; and a chapter from *Remaking Scarcity: From Capitalist Inefficiency to Economic Democracy*, a book written by one of us and discussing both Sahlins' classic essay and the connection between capitalism and consumerism. The point of these readings is to denaturalize the set of needs created by contemporary capitalism, showing how people's material needs always have to be analyzed in close connection with the social system in which they live.

Although from a chronological point of view it might seem to make sense to begin the session with Marshall Sahlins and his discussion of hunters and gatherers, the article on US millionaires is discussed first because, at first sight, it seems to confirm the neoclassical assumption of an insatiable human nature. The article reports on a Fidelity Investments survey of people who "had at least $1 million in investable assets, excluding any real estate or retirement accounts." The survey found that over 40 % of the people surveyed "said they did not 'feel wealthy'" and that many of them were worried that their wealth might not be enough to "fund their lifestyle" after they retired.

In opening the discussion, the author responsible for this section asked students how this article relates to the theme of the class, which is the meaning of being human. This question invites students to ponder whether this article has something to tell us about human nature. The researcher taught this article for a number of years now, but the surprising observation is how unsurprised students were by it. This was especially surprising, since the vast majority of City Tech students are from working-class or lower-middle class backgrounds and thus not from the ranks of millionaires accustomed to a lifestyle requiring exorbitant levels of wealth to sustain it. When students are asked why such a finding is to be expected, they usually give a mix of answers, ranging from claims regarding the insatiability of human nature to more socially situated claims regarding the effects of people's material insecurity as well as the influence of advertising.

The first type of claim allows me to highlight how pervasive and "commonsensical" the neoclassical "human insatiability" assumption seems to be. The second type of answer, on the other hand, helps to introduce the idea that human needs are socially constructed; in other words, people's attitudes toward material wealth are in many ways shaped by the nature of the social and economic system in which they live. This is an important insight that is completely missing from neoclassical economics, which tends to treat people's material needs and preferences as a black box. In the neoclassical model, people's material preferences are a pre-existing fact that is exogenous to economic life.[5] In other words, people's material wants are not seen as being co-determined by the economic system. Consumers are presented as sovereign, and free markets are viewed as their humble and efficient servant. Thus, the instrumentalization of human beings that lies at the basis of the thriving advertising and marketing industries is conveniently erased and capitalism appears as the benign force that, as Adam Smith[6] would have it, miraculously reconciles the pursuit of self-interest and profit with the common good.

After discussing the various aspects of capitalist society that prevent even millionaires from feeling rich, the researcher turned to Marshall Sahlins' classic essay. What makes this essay a perfect counterpoint to the Reuters article mentioned above is its explicit contrast of hunters and gatherers to the insatiable *homo economicus* postulated by neoclassical economics. Class discussion centers around the difference between the material desires of hunters and gatherers and those of contemporary millionaires. Students usually have no difficulty seeing that the desires of the former were more limited than those of the latter, so I encourage them to focus on the reasons for this difference. Consistent with my theme of denaturalizing human needs, the researcher jokingly ask students if their limited material desires make hunters and gatherers "perverts" who deviate from the human nature postulated by neoclassical economics. When they answer "no" with a smile, they are then asked how Sahlins' accounts for the hunters and gatherers' more limited desires. Thus, students are called upon to explain the link between the hunters and gatherers' material desires and their nomadic lifestyle, which is itself a product of the fact that, since they do not grow their food, they have to pick up and move whenever they deplete the food sources available in their immediate environment. Their nomadic lifestyle makes material possessions literally a burden, so hunters and gatherers are not interested in the accumulation of material wealth.

Thus, the contrast between contemporary millionaires who feel poor (or, at least, not rich) and hunters and gatherers who, in Sahlins' description, represent the original affluent society because they do not desire more than they have makes it clear to students that human insatiability is not a self-evident truth but an ideology that naturalizes the futility of capitalist consumerism. In so doing, the session also encourages students to analyze ideas about what it means to be human not just in terms of truth and accuracy but also in terms of power and the social effects they produce. It suggests that an uncritical acceptance of received truths regarding society and human nature may not just lead to incorrect perceptions of reality, but also facilitate the reproduction of social orders that may be oppressive and inimical to human well-being. Thus, the discussion of the human insatiability postulate not only involves students in a collective process of thinking critically; it also underlines to them why thinking critically about society and human affairs is so essential.

CASE STUDY II: QUESTIONING NEOCLASSICAL "RATIONALITY"

The second part of the session, entitled *The Near-Depression: The 2008 Financial Crisis and How It Happened*, begins with a critical analysis of the neoclassical assumptions about human behavior in the context of the workings of the USA and global financial system at the height of the housing bubble, high-risk mortgage lending, and other unregulated activities that preceded the crisis. Students consider the notion that, perhaps in retrospect, many of these assumptions would be somewhat obsolete in the context of twenty-first century market economies in light of the many regulations imposed since the Great Depression—the last major crisis that hit the US economy leading to a collapse of its banking and financial systems. However, a central focus of the case study is to bring to light the fact that, in practice, little had really changed, as banks, investors, mortgage lenders, and a host of other key players indeed acted upon the assumption that fewer regulations lead to more efficient markets; hence maximizing one's own self-interest is the most effective route to economic prosperity and originating mortgage loans to borrowers regardless of their ability to repay was good economic policy. A summary of the deregulation of banking and financial practices since the early 1980s provides the backdrop for students to understand some of the conditions that made such actions possible while providing a real world context in which students are encouraged to question the neoclassical assumptions underlying the concept of *homo economicus* and the free-market policy prescriptions upon which this model is based.

The major goal here is to challenge students to re-think each of the assumptions about human behavior in the context of the motivating factors that often shape the human capacity for effective judgment within the competitive capitalist economy. The case study itself begins with a pre-case study assignment outlining the learning objectives, a list of key terms, and a brief summary of the neoclassical assumptions about the motivations that, in a capitalist economy, guide the action of economic agents, such as individuals, business, and government. Prior to the class session, students complete two short readings[7] which introduce these assumptions. In particular, a 2009 article by the economist Paul Krugman posits the question of how so many economists could have missed the clear warning signs of the brewing crisis, while an excerpt from an article entitled "Neoclassical Economics" by authors Brennan and Moehler provide a theoretical grounding for the neoclassical assumptions about economic behavior. The central purpose is to prepare students (many of whom may not have previously taken an economics course) for a fuller discussion during the class session. In the preliminary reading assignment before class, students are first introduced to a summary of the neoclassical assumptions about human behavior in competitive market economies in the excerpt from the Brennan and Moehler article. They are then assigned the reading from Paul Krugman, "How Did Economists Get it so Wrong?"

Learning objectives focus on students' developing a critical understanding of

- The assumptions about human behavior in neoclassical theory as inherently rational, and the broad acceptance of the underlying assumption in mainstream economic theory that economic decision making is rational.
- How the assumption that consumers and businesses act as rational decision makers is built into the way economic participants are expected to behave in consumer-oriented capitalist economies such as our own?
- The question of whether human greed is a natural tendency that drives behavior, or is what is widely accepted as "rational" simply a way to justify greed.
- How the institutionalized acceptance of the theory of human rationality often informs behavior and actions in business and financial markets in decision making?
- Whether there are consequences to the unquestioning acceptance of the argument that pursuit of rational self-interest in a market economy always leads to the best outcome.

The class begins by asking students to identify the central arguments made by Paul Krugman and to interpret the more detailed neoclassical behavioral assumptions presented in the Brennan and Moehler excerpt. To elicit further debate, the author then challenges them to think about how these assumptions relate to the Krugman reading. Do any of the assumptions they were introduced to at the outset seem to be contradicted at all when placed in the context of different perspectives offered in the reading? Do the concepts of *rationality*, *efficiency*, and *scarcity* now take on different meanings in the context of the real life crisis discussed by Krugman? If so, then how? This discussion provides the setting for a short documentary film, *Meltdown: The Men Who Crashed the World*, which encapsulates the events leading to the near global financial meltdown and encourages students to think critically about these widely held assumptions regarding human behavior in the context of a real world crisis.

The film's documentation of the behavior of subprime lenders, banks, mortgage brokers, investors, Wall Street, the Federal Reserve chair at the time (Alan Greenspan), and government regulators at the height of the housing boom vividly illustrates how the concept of "rational" decision making in the idealized neoclassical sense became distorted by the motivation to "maximize one's self-interest." In many respects, the presentation shows how these two paradigms of the free market actually came into direct conflict with one another. The goal of individual utility maximization essentially clashed with the same goal at the organizational level as the actions of individuals ultimately contributed to the collapse of their own firms. Thus, "rational" behavior for the individual, say in the quest for making the most money from a financial transaction is shown to be at odds with the "rational" goals of the firm (e.g., the investment bank or the mortgage lending firm) to maximize profits and earnings while, at the same time, the drive to advance individual self-interest in fact undermines the goal of efficiency. One comes to the unavoidable conclusion that the ultimate outcome of self-interested behavior led to financial chaos and the near-collapse of the global banking system—the very antithesis of an efficient economic outcome. Scarcity, as it turns out, was actually quite relevant in this scenario. In its wake, the crisis produced mass unemployment, the loss of trillions of dollars of global wealth, and millions of foreclosures in the USA alone. Clearly, income, homes, and jobs became scarce very quickly.

As students view the film, which details the motivations and actions of financial institutions and many of their key decision makers, they are encouraged to think about how neoclassical assumptions about natural human behavior can effectively be deconstructed and questioned and to make note of key pieces of information that may seem to directly test these assumptions.

The film introduces students to some of the critical factors that provided the conditions for the financial crisis, including the growth of easy lending practices, a housing construction boom, and the rapid expansion of a relatively new type of home finance in the form of "subprime" lending, all within the context of a financial regulatory system that had been steadily weakened over the previous 20 years. By the early 2000s, a lending frenzy had taken off, with little concern about borrowers' ability to repay. In fact, ability to repay clearly was not the motivation behind loans made to borrowers with sketchy credit, few assets, and no money down. According to one California-based real estate agent during 2004–2005, "They [lenders] didn't really know or care about the qualifications of the buyers and whether people could make these payments or not wasn't much of a concern. If you could fog a mirror, you could get a loan."[8] Subprime lenders appealed directly to people with poor credit, while banks and mortgage brokers indulged in overtly fraudulent activities to "pump up" their mortgage business, offering complex loans with terms often hidden from borrowers. Many of these loans typically came with very low "teaser" rates that would re-set to much higher interest rates after just a few months. Poor and minority communities were major targets for such lending, a clear violation of anti-predatory lending regulations.

As the film reveals, there were many individuals seeking to "maximize their own utility" at the time. One such example spotlighted by the film was Angelo Mozillo, the head of Countrywide Financial, widely named the "undisputed king of the U.S. subprime market," who, at the height of the subprime boom during the early to mid-2000s, was reported to be earning an estimated $100 million per year. The problem with the notion that such utility maximization is necessarily "rational" is conspicuously evident in Mozillo's reflections, revealed post-crash, about the shoddy quality of the subprime loans he and his firm promoted. He reportedly had privately written that "In all my years in the business I have never seen a more toxic product," while simultaneously reassuring his investors and clients that everything was fine, stating "Countrywide views the product as a sound investment for our bank and a sound financial management tool for consumers."[9] This seems to clearly illustrate a divergence between motivations that may be rational for the individual from what would likely be rational for the larger good—in this case, the long-term profitability of the bank and, more importantly, the overall economy. The idea that 'rational utility maximizing' behavior that benefits the individual while ostensibly undermining the profit-maximizing goals of the institution comes to light in the wake of the crisis as Congressional and federal regulatory inquiries uncovered the inner workings of the subprime market and those responsible for the excessive risk taking that led to near financial collapse. Why did subprime lenders not care whether borrowers could repay their loans? Essentially, these loans did not remain on lenders' books. They were quickly bundled with similar loans from across the country, sold to investment firms where they were packaged into more complex "financial products" to meet growing demand from bankers for these high-yielding investments. Since there was virtually no regulation of the subprime market at the time, there was little risk to the various individuals and institutions that processed them along the way. Then chairman of the Congressional Financial Crisis Inquiry Commission, Phil Angelides, noted that, at every point in the process, from the broker to the lender, the securitizer, or the market maker, "everyone seems to have taken the view that they had no responsibility for the product that they were moving along in the system."[10] This plainly suggests that rational utility maximizing behavior for the individual was the central goal. As author William Cohan stated, "anybody who touched a mortgage made money,"[11] although there was little concern for doing so in the interests of the overall society (or even for the firms whose interests these individuals were supposedly looking out for). Following the film, the challenges and seemingly contradictory actions of the many players involved in creating the conditions for the crisis provide a context for the discussion that follows. The key questions guiding the discussion return to the neoclassical behavioral assumptions introduced in the pre-case study readings which are now viewed in the light of what the film reveals about the unquestionably destructive actions of individuals from mortgage brokers to bank CEOs.

Among the questions students are challenged to debate are the following:

1. Is there a fundamental inconsistency between the pursuit of individual self-interest and the pursuit of what is best for larger institutions (businesses, corporations, banks, etc.)?
2. Can the quest for rational (i.e., profit maximizing) behavior which involves taking actions that may inevitably lead to the collapse of the

business and the overall economy still be deemed as "rational" in this larger context?

3. Having considered these questions, is the attainment of an efficient outcome based upon supposedly perfectly informed decisions even possible?

4. What does it actually mean to be "perfectly informed"? If we accept the literal definition implicit in neoclassical theory as having all possible information to make an enlightened decision, would this not by definition include knowledge of all the possible repercussions and risks? If so, then were the decisions in this case truly perfectly informed or were they motivated by considerations of maximizing self-interest at the expense of all else?

5. Returning to the assumption that unimpeded competition results in the most efficient distribution of scarce resources, is this a realistic expectation given the reality that those responsible for key decisions at all stages of the process were aware that their actions in the quest for competitive advantage might not result in an efficient outcome for their firm or for the economy as a whole?

6. Given the concept of scarcity as a human-created construct in this case, a condition that resulted from the deliberate actions of those involved in creating the conditions for near economic collapse, how can economic conditions be manipulated to create scarcity in the wake of the crisis in the form of lost jobs and homes and a deliberate redistribution of income?

7. Finally, how can the theory of unimpeded free competition inform the question of whether regulatory constraints, if properly enforced, might have ensured a more efficient outcome for the overall economy?

Hence the goal is to engage students in thinking about the applications and relevance of a theory to real world events. The financial sector has consistently resisted and sought reduced regulations on its activities adhering to free-market beliefs. Thus, the concepts at the heart of traditional neoclassical theory examined in this case study appear to quite fittingly apply to the conduct of the various parties responsible for creating the conditions for the financial crisis.

In returning to the article by Paul Krugman, "How Did Economists Get it so Wrong?," students examine Krugman's argument that most economists' adherence on some level to free-market economic theory obscured their ability to recognize the presence of a housing bubble, the brewing subprime mortgage default crisis, and the obvious failure of government regulators to intervene before a full-blown crisis was underway. In Krugman's words:

> Few economists saw our current crisis coming, but this predictive failure was the least of the field's problems. More important was the profession's blindness to the very possibility of catastrophic failures in a market economy.... Meanwhile, macroeconomists were divided in their views. But the main division was between those who insisted that free-market economies never go astray and those who believed that economies may stray now and then but that any major deviations from the path of prosperity could and would be corrected by the all-powerful Fed. (2009, 1)

In other words, over time, the differences between economists' theoretical perspectives began to converge in many respects as they celebrated what former Federal Reserve chair Ben Bernanke termed "the Great Moderation," a nearly 20-year period from the mid-1980s through the mid-2000s during which recessions were mild and there appeared to be little need for major government intervention to control high inflation and unemployment. Krugman aptly quotes economist Robert Lucas, who proclaimed that the "central problem of depression-prevention has been solved."

Thus, the neoclassical theoretical assumptions at the center of the case study become profoundly relevant when viewed in the context of prevailing economic beliefs in the years preceding the crisis. The sense of complacency that took hold during the years of moderate business cycles persuaded many that the challenges posed by the Great Depression and unregulated capitalism were history. In Krugman's words:

> Until the Great Depression, most economists clung to a vision of capitalism as a perfect or nearly perfect system. That vision wasn't sustainable in the face of mass unemployment, but as memories of the Depression faded, economists fell back in love with the old, idealized vision of an economy in which rational individuals interact in perfect markets, this time gussied up with fancy equations. (2009, 1)

Understanding that many economists' moving back toward an acceptance of the traditional assumptions about the workings of markets and human behavior within them enables students to grasp the seemingly incomprehensible reality that so many failed to recognize the warning signs until it was too late. It also serves to bring the discussion back to the assumptions themselves with the objective of creating a more profound understanding of just how flawed this "idealized" world is, and why capitalist economies, and especially the financial sector, cannot survive and function without extensive government regulation. In the case study's written assignment, students consider the same questions discussed in class and evaluate how Krugman's arguments can be understood anew in the context of what they have learned about the financial crisis and the actions of individuals and institutions in facilitating it. The goal is to encourage a more informed perspective on the overriding motivations for human behavior in capitalist economies.

A final evaluation seeks to examine students' understanding of the previously reviewed assumptions about human economic behavior, and to assess a sense of new perspectives gained following discussion of the readings and the film in class. A two-page essay asks students to reflect on a few central questions:

> Based upon the discussion in the Brennan and Moehler excerpt, briefly summarize the central beliefs about human behavior as characterized by neoclassical economic theory. 2. Identify and discuss what Paul Krugman views as the flawed assumptions about human behavior according to traditional economics? How are these flaws seen as contributing to the failure to see the warning signs of the 2008 financial crisis? What does Krugman believe the discipline of economics needs to recognize and change in order to more effectively anticipate real world economic events? What do you view as the implications of Krugman's assessment for the neoclassical economic assumptions about natural human behavior? (2010, 946–951)

CONCLUSION

These two case studies, questioning human insatiability and neoclassical rationality, explore the question of what it means to be human in the context of an economic system that seeks to condition and shape human economic behavior for the purpose of perpetuating the existence and survival of that same system. Our goals at the outset were to encourage students to question some of the assumptions about what constitutes "human nature" and to consider the perspective that perhaps much of what has been unquestionably accepted by many as "natural" is actually cultivated.

At the same time, our objective was to foster a re-thinking of many of the neoclassical economic assumptions about what constitutes "rational" behavior in the context of real world events—in this case, the dynamics that led to the global financial crisis in 2008. Here, students were challenged to evaluate the major assumptions about human behavior in the context of the motivating factors that so often shape the human capacity for effective judgment within competitive capitalist economies.

Following a review of the neoclassical assumptions about economic behavior and obtaining a sense of how students understand and interpret these assumptions (a summary of which students have read prior to class) the documentary film, *The Men Who Crashed the World* is shown. Students are asked to identify two of three events or points from the film that made an impression. This usually leads to mention of the corruption and risk taking in the housing markets and financial markets that precipitated the crisis. The question of whether the decision of bank CEOs to market risky loans to borrowers whom they knew would not be able to repay and then to sell these loans as solid investments to investors would be considered "rational" elicits a range of responses. In one sense, students view these actions as a "rational" pursuit of profit and as motivated by a desire to maximize one's "self-interest," a conclusion many would unquestionably draw. However, when pressed further to consider the notion of rationality in the sense of what these individuals' actions meant for the larger economy or even the firms for whom they worked, many students reflect on the interpretation of rationality in other contexts. This generates questions such as "Why would a bank CEO (such as Angelo Mozillo) deliberately lead his/her profitable enterprise to the brink of collapse," "In what ways can this be seen as rational and efficient if the goal of private enterprise is profit maximization," and "Why would a lender make a mortgage loan to someone who is not asked for proof of income or who is not even expected to repay," or "Is this a quest for the best outcome possible"?

These questions flow into a discussion of how such actions can be reconciled with what we have come to consider "rational" and "efficient" according to orthodox economic theory. In the process, as students become engaged in a vigorous debate around these questions, they effectively participate in an important critical thinking exercise that fosters the consideration of other perspectives. At the same time, couching this critical thought in the framework of an issue that had far-reaching impacts on millions of people who lost their homes or jobs and that nearly led to a full-blown depression, the issue takes on new meaning for students. It also emphasizes for them why thinking critically about real world events is such an important part of being human.

Both case studies have sought to facilitate critical questioning and re-thinking of some widely held beliefs about human nature and what motivates human behavior under the social and economic environment characteristic of a consumer-centered competitive society. In doing so, both case studies introduced challenging concepts that ask students to consider—perhaps for the first time—alternative perspectives.

The concept of an insatiable human nature is investigated through a case study that enables students to critically evaluate this perception by considering two very different views about the fundamental motivations for human pursuit. The two readings—the Sahlins' essay that documents the hunter-gatherer society where the acquisition of possessions is cumbersome and encumbering to survival and the Kearney article that reveals the perception among many in present day society that the acquisition of more wealth is necessary in order to endure—draw very pointed contrasts about what "human nature" is. By reflecting on these contrasts, students are encouraged to question what is really "natural" about human life, while enabling them to see that insatiable wants have been conditioned by the type of society in which we live.

The second case study introduces further assumptions about human behavior in society that are rooted in traditional neoclassical economic theory. Here, "natural" human behavior is centered around the notion that economic "agents" seek to maximize their own self-interest, that such motivation is "rational," and that competition results in the most "efficient" distribution of scarce resources. As students become acquainted with these terms and their meaning in the context of orthodox economic principles, they are introduced to a 40-minute documentary film that reveals the decision of many individuals, in the run-up to the 2008 financial crisis, who were fundamentally not rational or efficient from the perspective of their impact on the national and global economies. They then consider Paul Krugman's scathing critique of most economists' failure to recognize the rapidly unraveling housing and financial system, which reveals how so many economists still cling to idealized conceptions of "natural" human behavior.

In short, these two case studies question the assumption that human behavior is determined by an unchanging human nature. Instead, they encourage students to recognize 'human nature' as a social construct. In modern capitalist societies human nature is constructed in line with capitalism's need for ever-growing levels of consumption. This means that abstract economic models which attribute to human nature behaviors and traits that are in reality produced by the prevailing socio-economic system are guilty of legitimizing rather than describing reality as it is.

NOTES

1. See William S. Jevons, *The Theory of Political Economy* (London: Macmillan, 1871); Alfred Marshall, *Principles of Economics.* (London: Macmillan, 1920); Leon Walras, *Elements of Pure Economics,* trans. William Jaffe (Homewood, IL: Richard D. Irwin, Inc., 1954).

2. In addressing the critiques, principally of non-economists at the time—that choice theory based on the assumption of rationality is flawed—Michael and Becker (1976), for instance, acknowledge that accumulating and investing in obtaining information can be costly, and as such "it is difficult to distinguish operationally between irrational choices and poorly informed ones" (1973).

3. Keynesian economics is named for the British economist John Maynard Keynes, whose theoretical perspectives on the fundamental flaws inherent in market economies gained wide recognition during the era of the Great Depression and had a profound impact on shaping public policy during the Roosevelt presidency. At its basis, his model of the macroeconomy demonstrated that government at the time was the only source capable of stimulating the USA (and other Western economies) back to a full employment level of output, given the exceptionally high rate of unemployment, and the collapse of both business investment and exports.

4. "Reagonomics" came to refer to the economic policies of President Ronald Reagan (1981–1989) who advocated widespread tax cuts and slashing social spending, along with increased military spending in the belief that such policies would create jobs and restore economic growth.

5. In this section the terms material "wants," "preferences," and "needs" are used interchangeably. The philosophical debate between "wants" and "needs" lies beyond the scope of this chapter. The closer the class session on which this section is based gets to this debate is when it discusses social scientific literature that shows human happiness to be more dependent on such factors as free time and the quality of one's relationships with other human beings than on growing levels of material consumption.

6. Adam Smith was a late-eighteenth-century economist perhaps best known for his work, *The Wealth of Nations.* Writing at a time when capitalism was an emerging new economic system, comprised mostly smaller enterprises, he postulated that the interaction between buyers and sellers in markets naturally found their "equilibrium" a point where both buyers and sellers agreed on a price for goods, a process that occurred naturally.

7. Paul Krugman, "How Did Economists Get it so Wrong?" and a brief two-page excerpt from the article, *Neoclassical Economics*, by Geoffrey Brennan and Michael Moehler.
8. Jim Kling in *Meltdown: The Men Who Crashed the World*, 2010.
9. From *Meltdown: The Men Who Crashed the World*, 2010.
10. Phil Angelides, Chair, Congressional Financial Crisis Inquiry Commission, *The Men Who Crashed the World*, 2010.
11. William Cohan, *House of Cards: A Tale of Hubris and Wretched Excess on Wall Street*, from interview in *The Men Who Crashed the World*, 2010.

BIBLIOGRAPHY

Aljazeera. 2010. Meltdown: The Secret History of the Global Financial Collapse. Directed by Terence McKenna. 2010. Toronto, ON: Canadian Broadcasting Corporation, 2010. DVD.

Becker, Gary. 1978. *The Economic Approach to Human Behavior*. Chicago: University of Chicago Press.

Brennan, Geoffrey, and Michael Moehler. 2010. *Encyclopedia of Political Theory*. Thousand Oaks, CA: SAGE.

Cohan, William D. 2010. *House of Cards: A Tale of Hubris and Wretched Excess on Wall Street*. New York: Anchor Books.

Friedman, Milton, and Rose D. Friedman. 1990. *Free to Choose: A Personal Statement*. New York: Mariner Books.

Jevons, William S. 1871. *The Theory of Political Economy*. London: Macmillan.

Kearney, Helen. 2011. "U.S. Millionaires Say $7 Million Doesn't Make You Rich, Survey Says." *Huffington Post*, Accessed March 14, 2011. http://www.huffingtonpost.com/2011/03/14/us-millionaires-say-7-mil_n_835327.html

Krugman, Paul. 2009. "How Did Economists Get it so Wrong?" *New York Times*, Accessed September 2, 2009. http://www.nytimes.com/2009/09/06/magazine/06Economic-t.html?pagewanted=all

Marshall, Alfred. 1920. *Principles of Economics*. London: Macmillan.

Michael, Robert T., and Gary S. Becker. 1973. "On the New Theory of Consumer Behavior." *The Swedish Journal of Economics* 75 (4): 378–396.

Panayotakis, Costas. 2011. *Remaking Scarcity: From Capitalist Inefficiency to Economic Democracy*. London: Pluto Press.

Sahlins, Marshall. 1972. "The Original Affluent Society." In *Stone Age Economics*, 1–39. Chicago: Aldine Atherton.

Sraffa, Pierro. 1926. "The Law of Returns under Competitive Conditions." *Economic Journal* 36 (144): 535–550.

Walras, Leon. 1954. *Elements of Pure Economics*. Translated by William Jaffe. Homewood, IL: Richard D. Irwin.

CHAPTER 4

Inducing Application of Interdisciplinary Frameworks: Experiences from the Domains of Information Literacy and Responsible Conduct of Research

Anne E. Leonard and Jean E. Hillstrom

Abstract Constructivist frameworks for information literacy and research ethics can be developed and nurtured in the context of an interdisciplinary course. Using the frameworks of two disciplines, students went on an experiential journey in support of ethics foundations through guest lectures and active learning exercises. This study describes the development and content of the responsible conduct of research and information literacy modules and discusses the role of each in an interdisciplinary course. Learning goals for both modules were evaluated by examining student responses in a free-writing exercise at the end of the semester, concluding with a discussion of the structural and concept similarities and implications for generalization of skills.

Both authors contributed equally to this work.

A.E. Leonard
Ursula C. Schwerin Library, New York City College of Technology,
City University of New York, Brooklyn, NY, USA

J.E. Hillstrom (jhillstrom@citytech.cuny.edu ✉)
Department of Social Science, New York City College of Technology,
City University of New York, Brooklyn, NY, USA

© The Editor(s) (if applicable) and The Author(s) 2016
R.D. Lansiquot (ed.), *Interdisciplinary Pedagogy for STEM*,
DOI 10.1057/978-1-137-56745-1_4

Keywords Human subject research • Information literacy • Interdisciplinary learning • Library instruction • Responsible conduct of research • Research ethics

Constructivist frameworks for information literacy and research ethics can be developed and nurtured in the context of an interdisciplinary course. Students in the writing-intensive interdisciplinary course, Weird Science, ask the question, "What does it mean to be human?" from a range of science, technology, engineering, and mathematics (STEM) and non-STEM disciplinary perspectives. While the guest lecturers offer responses to this question from their disciplines, the students must synthesize the disciplinary frameworks and the new knowledge introduced in combination with the content from assigned readings and class discussions.

Professor Anne E. Leonard, information literacy librarian and subject specialist for the English department, developed an information literacy module with the goal of helping students develop their abilities to find and evaluate scholarly and other information sources for their research papers. Jean E. Hillstrom, psychology professor and an experienced Institutional Review Board chair, vice chair, and administrator, developed a module that illustrates the principles of responsible conduct of research and human subjects research. Using the lenses of constructivism and social constructivism, the construction and delivery of the modules in an interdisciplinary course are analyzed. The manner in which students approached the intersection of research ethics and information literacy are examined by reviewing written work, class discussions, and student feedback. Finally, the ways in which students strove to become more information literate while engaging in responsible conduct of research in context of interdisciplinarity are explored.

CONSTRUCTIVISM AND SOCIAL CONSTRUCTIVISM

Constructivism, a theory about how people learn, proposes that individuals construct their own understanding and knowledge about the world from experiences and reflecting on those experiences. When we experience something new, we first try to understand it from previous experiences and perhaps by modifying our ideas when our prior understanding is inadequate.[1] Jean Piaget heavily influenced the direction of modern constructivist theory. He believed that people are active agents in their own development versus passive recipients of knowledge. He studied his own children and, in general, became particularly interested in children's wrong answers. He proposed that knowledge is constructed gradually as individuals interact with their environments in a process called adaptation. Individuals take in information from the environment and incorporate it into existing knowledge structures or schemas (assimilation). Since most schemas are inadequate to handle all new information, disequilibrium occurs, and individuals gradually adjust or create new schemas to make better sense of the information (accommodation). Equilibration is the process by which assimilation and accommodation are kept in balance.[2] Like Piaget, constructivism in the classroom acknowledges that students are active agents in their own learning. As Brooks noted, the teacher functions more as a facilitator who coaches, mediates, prompts, and helps students develop and assess their understanding, and thereby their learning. One of the teacher's biggest jobs becomes asking good questions.[3] Active teaching techniques—such as demonstrations, brainstorming, conducting experiments, role-playing, peer-review, real-world problem solving, and so forth—are examples of constructivism in the classroom.

Social constructivism emphasizes that knowledge is constructed through activity in a shared sociocultural context and in interacting with the environment.[4] Vygotsky's work strongly influenced social constructivists.[5] His sociohistorical theory of learning proposed that learning occurs through social interactions between more skilled or knowledgeable individuals and learners in the form of guided participation. Guided participation occurs when the more skilled or knowledgeable partner actively engages the learner in the zone of proximal development (the difference between what the learner can do alone versus with assistance). Thus, it is important for the instructor to have a sense of students' current knowledge and ideas so that they can encourage students to engage in thoughtful assimilation and accommodation of new information. For example, on the occasions where a student has plagiarized in a paper, the student may be required to meet with the instructor in order to discuss the issue. In a discussion as to what plagiarism is, students more often than not state, "It was on the internet—I thought it was free to use," or something similar. This interaction presents an opportunity to refine the student's current knowledge by expanding his or her understanding of what exactly plagiarism is and that information from the internet still needs to be paraphrased or quoted and cited appropriately. In students' later work, typically there is significant improvement in how information from the internet is used and documented.

RESPONSIBLE CONDUCT OF RESEARCH MODULE
AND LEARNING GOALS

The research ethics module is typically scheduled about one-third of the way into the course—after the philosophy and biology guest lectures and a discussion of the Henrietta Lacks[6] case. Other course lessons delve into research ethics, including plagiarism (during the introduction of the course in the first class meeting and reminders throughout the course), the Henrietta Lacks case, and the information literacy module.

The learning goals of this module expanded on the course's interdisciplinary goals: they provide and engage students with an understanding of ideas and connections in the natural sciences, social sciences, technology, and engineering, including the cultural factors that affect these disciplines, as well as considering philosophical, historical, and ethical perspectives. The learning goals for the research ethics module include the following:

- Develop an understanding of what is involved in research ethics;
- Appreciate the need for ethics in research;
- Experience the challenges of determining what is ethical in research;
- Recognize ethical and unethical research;
- Identify ethical issues of failing to obtain informed consent;
- Generate alternatives to unethical research that would be considered ethical; and
- Practice ethical research standards.

As part of their homework assignment for this module, students complete the Responsible Conduct of Research online course (www.citiprogram.org), which is offered by the Collaborative Institutional Training Initiative (CITI Program) at the University of Miami. In a short, free-writing exercise at the end of the semester, two questions pertinent to this module were asked: "What does research ethics mean to you?" and "What are the five most important ethical issues that students scholars/researchers need to know about as they conduct their research?" Results are presented later.

In the class presentation, responsible conduct of research is defined as multifaceted and as involving the application of fundamental ethical principles to animal and human research, in addition to the more familiar academic domains, such as plagiarism and fabrication of data. Even so, students often interpret research ethics as avoiding plagiarism and probably do not consider the broader implications or potential personal relevance.[7] When presented with the problem of research ethics, instructors often encounter students with an attitude of "Who cares?" or "This doesn't apply to me." So, to induce personal relevance, the immediacy of the problem is highlighted with a discussion of the infamous Wakefield autism study that led to hundreds of parents not obtaining vaccines for their children. This study "hooks" students into the topic because of its currency and because most students have heard about the supposed link between vaccines and autism. However, any recent study that illustrates violations of research ethics principles or even a vivid historical example (e.g., the Stanford Prison Study, which is discussed later in this section) could be used. In summary, Dr. Andrew Wakefield and colleagues published a study in 1988 that linked the measles, mumps, and rubella (MMR) vaccine to autism. Numerous attempts to replicate his findings failed to find evidence that the MMR vaccine caused autism in young children. Despite the lack of evidence, beliefs persisted and the number of new cases of measles steadily rose in the USA and Europe. Although the Wakefield study was resoundingly refuted on a number of grounds, including the falsification of data by the authors, the effects on public perception still persist and have the potential of affecting everyone's daily lives.[8] For example, the spread of measles originating from early 2015 California Disneyland outbreak was attributed to the fact that a substantial number of children in the USA that were not vaccinated.[9]

Various historical case studies are presented, illustrating ethical concerns. Students are then asked the following questions: (1) what was the objective of the study? (2) was the research conducted ethically and why/why not? (3) could this study have been conducted in a more ethical way? Finally, when discussing research ethics, it is noted that there are often no "right" answers, and that the journey may be as important, if not more so, than a final decision regarding ethical treatment of subjects in research.

Throughout the discussion, Ruben Bolling's "Tom the Dancing Bug" cartoon is used as a tongue-in-cheek rule to approach research ethics from different perspectives while examining personal biases. The cartoon, presented as a chart, is a thought-provoking guide for moral behavior based on the premise, "The more a living being is like you, the nicer you must be to it."[10] Various categories are listed with examples in a hierarchy ranging from plants (e.g., radish) at the lowest level to immediate family members (e.g., daughter) at the highest level and asks how much the category/example is like you (the reader). Each category is paired with an "appropriate

moral response" along with a guide to answer the following four questions: "Should you help it?"; "Can you harm it?"; "Can you kill it?"; and
"Can you eat it?"

For instance, in the category, pets/primates, the example is a dog. In
response to the question of how much the dog is like you, the answer is
"Not human, but anthropomorphized." In response to the four questions, the guide indicates that you can help it, "If you're in the mood";
sometimes you can harm it, "Depending on circumstances" (such as for
research); and sometimes, you can kill it, "Depending on circumstances";
and no, you can never eat it. Throughout the research ethics module,
this guide was used to facilitate examining personal assumptions about
research ethics, whether human or animal, as explored in the historical
case studies.

Animal research ethics is discussed next to contrast with human subjects research ethics. Before the Animal Care Act was enacted in 1966,
researchers and individual labs were responsible for ethical care of animals
used for research purposes. An amendment four years later made provi-
sions for animal use and care committees (now known as IACUCs) to
oversee animal research at institutions. Harlow's research was presented
because of his careful experimentation, the ability to compare his animal
studies with human studies, and because baby monkeys are often perceived as "cute" and elicit anthropomorphism. There are online video
clips of archival footage where Harlow explains his research, but images
could also be presented via slides. Harry Harlow is probably best known
for his research on attachment using rhesus monkeys. In his basic research
paradigm, newborn monkeys were removed from their mothers and
placed in a cage with a wire mother and a cloth mother (the wire mother
covered with fabric). Half of the monkeys were fed (a bottle was inserted
in the mother's frame) by the wire mother and the other half by the cloth
mother. Results showed that the monkeys spent significantly more time
on the cloth mother regardless of which mother fed them. Harlow concluded that attachment in the monkeys was based on contact comfort as
opposed to whom fed them; he generalized these results to attachment in
humans.[11]

Harlow's animal research was a lesson in contrasts. On one hand,
his research illustrated the emotional lives of his animal subjects (e.g.,
attachment, effects of social isolation) in direct contrast to the dominant
mechanistic behavioral theories at the time. But his research has also been
criticized for its inhumane treatment of animals—ill effects included brain

lesioning, maternal deprivation, social deprivation, and "pits of despair" isolation chambers.[12]

Students in the most recent section of the class generally had no difficulty identifying the purpose of the Harlow's study, recognizing that he studied animals and not humans because it would be clearly unethical to remove human babies from their mothers to test attachment theories. In Tom the Dancing Bug, the relevant category is pets and primates. Students expressed various responses, ranging from research on animals is OK if the animals are protected, to we should never experiment on animals. Most students had difficulty identifying more ethical ways to conduct this research, perhaps because of their unfamiliarity with animal research protocols. Some students were surprised that, despite the ethical issues of Harlow's research, a similar research study was recently approved.[13]

The next case study presented was the Tuskegee syphilis study because of its seminal influence on legislation in the USA and that it clearly violated a number of ethical imperatives. Further, this study helps illustrate timelines for human and animal research legislation. Archival photos were shown via PowerPoint, as well as pictures of individuals with syphilis. In the Tuskegee study, in 1932 The US Public Health Service began investigating the transmission patterns and long-term outcomes of syphilis, a poorly understood disease at that time. The subjects were 600 poor, illiterate black men in rural Alabama. The men were never informed that they had syphilis—they were just told that they had "bad blood." In return for their participation, they were given free medical care, food, free burial insurance, and a certificate after 25 years of service. In 1972, 30 years after penicillin was established as a treatment, the Tuskegee syphilis study was brought to light by an investigative reporter and led to the Belmont report articulating basic ethical principles for the protection of human subjects in biomedical research, as well as a number of reforms.[14] Again, students had no difficulty identifying the Tuskegee study's objectives and were clearly able to identify that the study was unethical on a number of grounds, including lack of informed consent and harm to subjects (particularly since penicillin could treat the disease). Students generated a number of solutions to the ethical problems with the study, including fully informing the subjects about the research and stopping the study when syphilis became treatable. However, in using the cartoon morality guide, students had more difficulty identifying the category that would represent poor, illiterate, black sharecroppers. Student responses seemed to center around the community member category, but sometimes noting the outsider category might apply, perhaps because of perceived differences between the subjects in the study and themselves. Students also had difficulty with the "appropriate moral response" presented by the cartoonist for the categories but were able to generate more ethical methods as noted earlier. With regards to timeline, it was pointed out that at approximately the midpoint of the Tuskegee study, the Nuremberg Code came into being in 1948 because of the atrocities committed in medical research on prisoners during World War II. The Code made it clear that voluntary, informed consent in research is essential; the benefits must outweigh the risks and ten other points with regards to medical research.[15]

Research ethics in behavioral studies was not directly addressed until the National Research Act of 1974. Two vivid behavioral studies were chosen to illustrate ethical concerns: Milgram's study on obedience and Zimbardo's study on role-playing. For both examples, online video clips from YouTube presented archival footage. In the early 1960s, Stanley Milgram designed a behavioral experiment to see if people would obey authority figures even when the instructions were morally wrong.[16] He presented the study as a memory experiment, deceiving the subjects as to its true purpose. The basic design of the study was as follows: two subjects arrive to the study and wait in the waiting room. The researcher "randomly" selects which subject will be the learner and which will be the teacher. In actuality, the drawing is not random; the learner is a confederate who follows a specific script, and the true subject is the teacher. The researcher is present throughout the experiment. The teacher is seated in front of a machine that delivers increasing levels of electric shock to the learner. (The teacher receives a sample shock to show the machine is "real"; however, no shocks are actually delivered to the learner.) The teacher is instructed to read pairs of words and then test the memory of the learner. If the learner makes a mistake, the teacher is instructed to give a shock, increasing the intensity of the shock with each mistake. The learner follows a prescribed script that includes correct and incorrect responses, actions (e.g., pounding on the wall as the amount of electrical shock apparently becomes more and more intense), and finally silence. The experimenter also follows a script with prescribed prompts (e.g., "the experiment requires you to continue").[17] The variable of interest was how high of a level of shock the teachers would give before disobeying the experimenter (if they disobeyed at all). Milgram's results showed that 65 % of the subjects continued to the highest level of shock (450 volts, marked XXX on the equipment).

With regards to research ethics, it was pointed out that Milgram did seek the consent of his subjects and because he used deception, he debriefed them afterward. The subjects were also told beforehand that they would receive payment whether they completed the procedures or not. Observers noted that the subjects appeared to experience extreme stress during the procedures. However, Milgram expressed the view that there were no long-lasting effects of the stress experienced during his experimental procedure.[18] In response to the question regarding the objective of the study, students did identify that Milgram was studying obedience to authority, but also mentioned that perhaps certain people would be more likely to shock others (e.g., more aggressive individuals). The researcher's question was then posed to the students—"Would you obey the experimenter?" Results ranged from "Never" to "I don't know." Students seemed challenged in generating alternatives to Milgram's methodology, perhaps because none of the students were Psychology majors. Two recent studies were presented that replicated Milgram's work but with modifications to reduce subjects' stress. Both studies found similar results.[19] In applying the Tom the Dancing Bug framework, students seemed to center around the community member category.

In discussing Philip Zimbardo's famous behavioral study, the class looked more closely at the idea that the situation may determine a person's behavior. Zimbardo and colleagues created a realistic prison environment at Stanford University and set out to study the influence of a simulated prison setting and role-playing on interpersonal interactions as subjects took on the role of "prisoner" or "guard."[20] Potential subjects (male college students) consented to participate and were given psychological tests to eliminate the hypothesis that the subjects may have pre-existing dispositions toward aggressive or punitive behavior. The subjects randomly assigned to be a "guard" wore uniforms, worked 8-hour shifts, and were free to go about their business otherwise. According to Zimbardo, the guards were minimally instructed in how to do their jobs: "The 'guards' were free with certain limits to implement the procedures of induction into the prison setting and maintenance of custodial retention of the 'prisoners'."[21] The subjects randomly assigned to be prisoners remained in the prison setting for the duration of the experiment. They also received minimal instruction on what to expect in the prisoner role. Within a short period of time, subjects conformed to their roles, sometimes to the extreme. Although this study was originally planned for two weeks, it was stopped after only six days because of the greater than expected aggressive behavior of the prison guards and the mental and physical deterioration of the prisoners. Given that the subjects were randomly assigned to conditions, Zimbardo and colleagues interpreted results to mean that the situation determined the resulting behaviors, not characteristics of the individuals.

Students had more difficulty identifying the objective of the Stanford prison study, with responses ranging from experiencing prison life to conformity to obedience. They also had more difficulty in generating alternatives to the methodology that were more ethical. Responses to Tom the Dancing Bug were also more varied, but the number of students who discussed this study in their final research papers indicates that perhaps they identified more with subjects in the prison study because they were college students as well. It is also possible that popular media may have affected students' perceptions due to the fact that several movies have been made on the topic (e.g., *The Experiment*, 2010; *The Stanford Prison Experiment*, 2015) as well as a documentary (*Quiet Rage: The Stanford Prison Experiment* 1992). Zimbardo's conclusions were brought to bear on the questions students had earlier in discussion of Milgram's research, "Would you obey the experimenter?" Zimbardo would have concluded that the situation could be more powerful than individual characteristics. The Abu Ghraib scandal was also briefly discussed in this context.

After reviewing the historical cases, it was pointed out that the 1974 National Research Act provided for the creation of a committee to establish basic research protections for humans in medical and sociobehavioral research. The resulting Belmont Report established three fundamental principles in human subjects in biomedical and behavioral research: respect for persons (e.g., informed consent), beneficence (e.g., benefits outweigh the risks), and justice (e.g., fairness).

Teaching Research Ethics in an Interdisciplinary Course

Several questions arise when considering teaching research ethics: (1) why teach research ethics? (2) why teach research ethics in an interdisciplinary class? (3) is there a difference in purpose or outcome? As to the first question, several answers come to mind: research ethics training may be required; it should be a part of research education, ambiguities such as authorship and sharing of data often arise, and disciplines likely vary. At the undergraduate level, research ethics involving plagiarism, proper citation, and paraphrasing is often taught in courses where students learn to write papers, such as English composition courses. Unfortunately, students often do not generalize the skills learned in the composition classes to writing papers in, say, Psychology classes. Given this compartmentalization, it is not surprising that students may relegate avoiding plagiarism to English or similar classes, and regard issues of human and animal research ethics as only pertaining to research methods classes. Similarly, animal and human subjects research ethics is commonly discussed in disciplinary contexts such as introduction to Psychology or research methods courses, often remaining compartmentalized in these disciplines.

One of the challenges in teaching research ethics is fostering the generalization of critical thinking skills and ethics frameworks to bear on areas outside of a specific discipline. One way to combat the specialization of research ethics within a discipline is to teach responsible conduct of research in an interdisciplinary course. Thus, the overarching goal of the interdisciplinary Weird Science course was to bring disciplinary perspectives found in the natural and social sciences, technology, and engineering fields to bear on an interdisciplinary question of what it means to be human. The interdisciplinary course is a natural vehicle for fostering the application of knowledge and skills to solve real-world problems that cannot be addressed within a single discipline. The research ethics module employed a set of questions and a framework to foster ethical reasoning as students learned about, reasoned through, and examined their personal biases and assumptions while working through several vivid case studies from different domains. Further, focusing on ethical reasoning, as opposed to learning codes of ethics or rules to be followed, helps students generalize skills applicable to other disciplines and their own lives.

Assessing Responsible Conduct of Research Competency in an Interdisciplinary Course

Students' responses to two of the three questions posed at the end of the semester were analyzed using NVIVO version 10 software. NVIVO eliminates articles, conjunctions, and other sentence structure words (e.g., a, an, the, for, or, and, therefore) and has the ability to categorize redundancies such as plurals and past tense into root words (e.g., cited, citing, cited = cite) as well as generalize to include synonyms (e.g., information = data, information, source, sources). In responding to the question, "What does research ethics mean to you?" a word frequency analysis was conducted, and words were ranked according to weighted percentages (the frequency of the word relative to the total number of words). Aside from the question's stem words, "research" and "ethics" (two of the most common words), the most common words included "conducting," "study," "apply," "information," and "credit." These words appeared to center around conducting ethical research—one student spoke of "being mindful of how are you conducting your research"—and documenting information by crediting sources. For instance, a student commented, "If doing research on a paper and information from some source is used, then the ethical thing to do is to cite this source and give credit to where it came from."

A word frequency analysis was also conducted for the second question, "What are the five most important ethical issues that students scholars/ researchers need to know about as they conduct their research?," using the same parameters noted above. The most common words included "information" and related phrases, such as "informed consent," and "obtain relevant information"; "research," as in the phrase, "Is the researcher prepared to conduct the research responsibly …"; "subjects," such as in responses like "confidentiality with your subjects" and "topics"; "credit," as in the phrase, "give credit"; and "using," such as in the phrase, "using credible material." For several of the words, such as "information" and "subjects," the word frequency analysis pointed to both human subjects research topics (e.g., "informed consent," "confidentiality with your subjects") as well as information literacy topics (e.g., "obtain relevant information," "topics").

Students' responses to the questions were also coded in NVIVO to reflect thematic content derived from the general and specific learning goals from each module. The first two questions are discussed here. The themes in the general category of responsible conduct of research included recognizing unethical behavior, obtaining informed consent, doing no harm, acting ethically, being honest, and working within frameworks. The information literacy themes were using information appropriately, learning, maintaining the quality of information, citing correctly, identifying the need for information, finding relevant information, evaluating material critically, acting ethically, being honest, and working within frameworks. A few themes applied to both: ethics, honesty, and frameworks. Prior to the qualitative analyses, it was expected that the first question would result in proportionately more references to responsible conduct of research as presented in the research module. However, in reviewing the resulting themes, the responses were split between responsible conduct of research with 24 references, and 21 for information literacy. For the second question, a mix of references was expected between the two general categories, with more references in the category, responsible conduct of research. The results showed that more statements occurred for responsible conduct of research (50 references) than for information literacy (38 references), although the frequencies are still relatively close.

INFORMATION LITERACY MODULE AND LEARNING GOALS

A library presentation and active learning workshop has great practical value in any course that includes a research paper, but a typical "show 'em the databases" demonstration by a librarian does not adequately meet the information literacy competencies demanded of students in a writing-intensive interdisciplinary course. Instructional strategies that address interdisciplinary learning and support student research projects are grounded in information literacy principles. In Weird Science, students annotate assigned readings from a range of disciplines, perform a literature review, and write an annotated bibliography and a research paper, all activities that require the ability to research successfully. Assignments in interdisciplinary courses require research instruction and support from a librarian that is distinct from a librarian's demonstration of database searching to students. Simply demonstrating the use of a few databases is not productive, as commercial database search interfaces change rapidly and libraries continuously develop their collections of electronic resources, rendering search skills obsolete quickly. Since many library databases are oriented toward the literature in one subject, emphasizing search proficiency in just one or a few databases will not benefit students in an interdisciplinary course; they benefit more from an understanding of search strategies and critical information evaluation. The role of information literacy, as well as the role of the librarian in interdisciplinary learning, can be simply to facilitate students' efforts to locate relevant information sources from a range of disciplines. Yet as a framework for exploring a problem from multiple disciplinary perspectives, information literacy can be a means to enhanced interpretation of new information and assimilation into the context of students' lived experiences via a constructivist approach. Information literacy is critical to interdisciplinary learning settings where students encounter information from a range of disciplines, in assigned readings, in-class lectures and discussions, and on their own as they develop their literature reviews and research papers. An information literate student—one who is able to identify his or her need for information and use appropriate research tools to evaluate information for relevance, the expertise of the author, currency, accuracy, and purpose—is better able to assimilate and contextualize information from various sources, as well as applying basic principles of evaluation to determine the information sources that best meet his or her needs.

The course syllabus lists six course learning goals, including two especially pertinent to information literacy: methods for finding pertinent information and critical evaluation of ideas and their sources. These goals guided the development of learning goals for the information literacy session, that students successfully locate high-quality, relevant information resources; evaluate all information resources (especially internet sources) according to important criteria (the criteria were introduced by the instructor and refined by the students through discussion and voting); cite all sources correctly and consistently; and pick up some transferable knowledge that they could use in other courses with research assignments.

Over several semesters, the role of the librarian as a guest lecturer in Weird Science has evolved, depending on the timing of the presentation with respect to the students' progress on the research paper and literature review assignments. Typically, the one-shot library research skills class starts with a short lecture and demonstration of the use of library research tools to locate high-quality, peer-reviewed scholarly articles and the use of citation tools to cite and document articles. In Weird Science, the lecture-heavy approach evolved into a discussion on evaluating information, regardless of its source. The introduction of an information evaluating game, in which students competed in teams to find the highest-quality information source on a given topic, brought an active learning exercise to the class in order to reinforce the skills taught. The game afforded the opportunity to better assess students' understanding of the purpose of evaluation and their own evaluation skills.

The coordinator of information literacy at the New York City College of Technology (City Tech) library designs and teaches one-shot library research instruction classes at the request of the classroom instructor; for Weird Science, this involved collaborating with the classroom instructor to teach relevant research concepts and skills that assist students in the early stages of their research assignments. Designing a single 70- to 90-minute library lecture and workshop to teach all aspects of information literacy is challenging, and the essential information literacy competencies that are most congruent with the learning goals of the course must be identified and prioritized. An active learning classroom activity that gave students the opportunity to reinforce their research abilities through guided use of library subscription databases was preferred over a traditional lecture, with an overarching goal of locating a few sources potentially relevant to their research paper topics. A presentation, "The Good Stuff Is Out There: Finding and Evaluating Your Information Sources," foregrounded the ability to identify bias and analyze quality of authorship or sponsorship of information found on the internet or through searching library subscription databases. Students addressing the larger issue of interpreting and defining humanity surely encounter dubious, unreliable, and biased information sources, especially when searching "in the wild" online. The brief lecture on information evaluation focused on the application of RECAP criteria: Relevance, Expertise, Currency, Accuracy, and Purpose. This is the City Tech library's own version of the CRAAP test (Currency, Relevance, Accuracy, Authority, and Purpose) widely used by academic libraries. Instruction librarians at City Tech decided to replace Authority with Expertise, acknowledging that authority can be constructed differently depending on the knowledge community it springs from, and thus may be more effectively conceptualized as the creator's expertise on that particular topic for a particular audience in a specific information medium.[22] Along with an understanding of purpose, or the reason or motivation, for creating a particular information source and its intended audience, students become better prepared to evaluate information based on the appropriateness of its producer and the context of that source in relationship to similar information sources.[23]

The Evaluation Game

To develop an active learning activity for the information literacy module, learning goals for the course were consulted, which include the mastery of methods for finding pertinent information and the ability to evaluate critically ideas and their sources. The evaluation game use has its roots in an information evaluation game, Quality Counts, developed by a colleague.[24] This game (which was also used in a semester-long information literacy class) was adapted for Weird Science by shortening the gameplay time so teams of students could determine collectively evaluation criteria and play a few rounds within half of a class meeting period. After a brief lecture about the importance of evaluating information, in a discussion, students determined criteria to evaluate information. They generated the following list: website not sponsored, publication date, copyrighted, objectivity of website content, credibility of the web domain, grammar and syntax (writing mechanics), author's credentials, presence of a references list. Students then voted to narrow this list to the top three, and selected the credibility of the web domain, the author's credentials, and the presence of a references list.

Working in teams, the students competed to find two information sources that both offered an answer to the question, "What evidence is there that global climate change influenced the evolution of human beings?" and met the three criteria. The facilitator of the workshop and game scored the websites they chose, one point for each criterion met. At the end of gameplay, two teams with a tied score competed to see who could find the least credible website. Their selections were evaluated with the three criteria reversed: unreliable web domain, flimsy or ambiguous author credentials, and the absence of reference lists or links to data sources. Although not every team participated in the tie-breaker, engagement around the classroom was palpable and students instinctively located the humor in poor quality online information sources.

How Information Literacy Competency Enhances Interdisciplinary Learning

The information literacy librarian facilitated students' learning about knowledge creation in various disciplines. They learned to understand the importance of the literature review and the types of methodologies appropriate for a variety of research questions and disciplines. Since students learn about the use of library databases as well as the internet to find potential information sources for the research paper, the task of evaluating information, including understanding context, purpose, and the expertise of the producer, is especially important.[25]

The Association of College and Research Libraries (ACRL) Information Literacy Framework presents six core concepts that, once grasped, facilitate students' ability to comprehend disciplinary and interdisciplinary ways of thinking: (1) authority as constructed and contextual; (2) information creation as a process; (3) information has value; (4) research as inquiry; (5) scholarship as conversation; and (6) searching as strategic exploration.[26] Students' essential knowledge practices that reflect engagement with these concepts ideally include their understanding of the knowledge creation cycle and their appreciation of different research methodologies in different disciplines, as well as understanding how authority, or the relative importance of authorship, can shift depending on the venue, the purpose, or the intended audience of the information.[27] Students come to view scholarship as a conversation and demonstrate the ability to insert themselves into this conversation. Students understand—and practice—research as an iterative process, learning how to develop search strategies and refine keywords through the process of searching and re-searching. Interdisciplinary learning denotes knowing about knowledge production, consumption, organization, and application from a variety of disciplinary lenses.

Assessing Information Literacy Competency in an Interdisciplinary Course

The application of the new ACRL Framework's threshold concepts, especially research as a conversation, affords exploration of how academic authors incorporate multiple disciplinary lenses in their published writing. One way to achieve this is through an in-class demonstration of how an author of an assigned reading cites authors of other assigned readings, and an activity in which students use research databases to locate an article relevant to their topic, then use a citation database to see who cites and is cited with special attention to cross-disciplinary citing.

Another means to assess students' abilities to assimilate information from many disciplines is the use of visualizations to show relationships among important concepts. Dilevko and Solgasnova described the use of knowledge map creation in identifying emerging areas of knowledge.[28] Knowledge maps of the knowledge domains of recent dissertations demonstrate emerging areas of knowledge and foster a mindset of interdisciplinarity. They document how librarians' creation of knowledge maps of academic fields and subfields in recent dissertations represent visually a way that librarians can guide emerging scholars (doctoral students) in identifying relevant and recent literature in their subfields and areas. Similarly, Weird Science students developed concept maps to develop clear thesis statements for their research papers. Their concept maps visualized relationships between prominent concepts from class discussion and the themes of shared readings. By visualizing relationships between concepts and the readings that emphasize them, the concept maps illustrated important questions that researchers have addressed across disciplines and helped students narrow down a broad topic into something both manageable and relevant. Students' ability to identify important concepts and relationships between them was evident, making the concept map straightforward to assess.

A simple rubric to assess the references lists of students' literature reviews was used to evaluate students' ability to locate information sources appropriate for the assignment. The criteria used included whether the information was produced by credible authors, reasonably free of bias, current, and accurate. Below is the rubric: (Table 4.1).

For the most recent section of this interdisciplinary course, most students' references lists scored between three and four out of a possible four points; out of 20 References lists evaluated, five scored 4 points (equivalent to an A grade), five scored 3.5 points (A–/B+), eight scored 3 points (B grade), one scored 2.5 points (C grade), and one scored 1 point (failing grade). The information resources that students incorporated into their literature reviews generally indicated greater than average competence with finding scholarly articles and books via internet searching and library database use. Quite a few references lists contained citations to non-scholarly sources for which an expert author could not be established; those that included less reliable sources tended to refer to more than one such source, which resulted in a lower score. The references lists that scored the highest included one or fewer low-quality, less reliable sources. The average score was 3.25 out of four points, equivalent to 81 %, or a B-.

In another assessment near the end of the semester, students responded to these questions about their understanding of research ethics and information literacy:

1. In a few sentences… "What does research ethics mean to you?"
2. What are the five most important ethical issues that students, scholars/ researchers need to know about as they conduct their research?
3. In a few sentences… "What does information literacy mean to you?"

Table 4.1 References list assessment rubric: quality and appropriateness of information sources

One point	Two points	Three points	Four points
All or almost all sources used are not appropriate for the assignment, and contain inaccurate, biased, or outdated information from inexpert authors	Most sources used are not appropriate for the assignment, and contain inaccurate, biased, or outdated information from inexpert authors	Some sources used are not appropriate for the assignment, and contain inaccurate, biased, or outdated information from inexpert authors	All sources used are appropriate for the assignment. They are credible sources, and are accurate, expert, objective, and current

The responses were analyzed from an information literacy perspective for this module. A visualization of word frequency in their responses revealed some patterns. Suppressing the words "research" and "information" and accounting for duplicated word forms such as "issues" and "issue" led to clearer results. The words "ethics," "literacy," and "important" appeared most frequently; this indicates a connection between the concepts of research ethics and information literacy and the weight or importance assigned to those concepts. At the next tier were words that suggested an understanding of ethical research conduct with humans: "human," "being," and "students." The latter suggests that students connect the concept of ethical research conduct to specific issues or examples, which they learned about in the Responsible Conduct of Research guest lecture. A qualitative perspective on student responses to the question, "What does information literacy mean to you?," shows thematic patterns and commonalities. Most student responses addressing the definitions of information literacy can be grouped into a few dominant themes:

- Ability to identify one's own need for information;
- Ability to evaluate the quality and relevance of information in any medium or from any source; and
- Awareness of academic integrity and the consequences of plagiarism.

These themes demonstrate a complex understanding of information literacy as it applies to research, first recognizing that a researcher must identify his or her own need for information before attempting research. Students showed an awareness of the researcher's responsibility to evaluate information for quality and appropriateness regardless of the source, a key ability for lifelong learning beyond the undergraduate classroom. Such an awareness of academic integrity reflected discussion of this topic throughout the course, and is surely reinforced in other courses with significant research and writing components. Their awareness of the importance of information literacy competency as a means to better academic production was clearly shown through comments such as information literacy is "the ability to adequately discern important information ... and apply it to your focus of interest." Further, students' perception of the value of information evaluation was evident in comments like "understanding what references or sources are right for your research. For example using the *New York Times* versus a blog." One student identified information literacy as an essential life skill: Information literacy "is imperative for students to become independent lifelong learners. Information literacy provides the opportunity to equip us with critical thinking skills." A few responses contrasted true learning with mere fact memorization, stressing the understanding of theory and the application of facts. One response defined information literacy as the ability to do a close reading for comprehension of any content, any media, be it a research article, novel, or video. A few students appear to have turned to the internet to inform their definitions of information literacy. One student wrote, the "official meaning [of information literacy] is the ability to know when there is a need for information, to be able to identify, locate, and effectively use that information, … that meaning is the very essence of what it means to me also."

When verbs in students' responses to the question, "What does information literacy mean to you?" are grouped and counted, four information-using actions critical to successful research and evaluation of sources stand out: "evaluate," "identify need for information," "use ethically," and "find." They identified these tasks as the essential components of information literacy. Using varied assessment methods, including surveying students on their knowledge of research ethics and information literacy as well as evaluating student artifacts, allows for a more complete picture of students' abilities to assimilate ethical aspects of research than a single assessment method.

Looking ahead to future semesters and information literacy guest lectures in this course, more rigorous assessment of students' information literacy competencies would be valuable and best achieved through consultation with the course's creator and facilitator. Means to assess students' understanding of information literacy concepts and practices could include a pre-test or an in-class activity in advance of the information literacy module, a method shown to be effective by Natalle and Crowe.[29] Such a pre-test could be a simple questionnaire asking students to self-assess their ability to identify high-quality information sources appropriate for a research paper, or completion of an online tutorial and quiz in which students apply evaluation criteria to a variety of scholarly and popular sources. Another means would be a more exhaustive evaluation of students' research papers and literature reviews for ethical use of all information sources and correct, complete documentation of their sources. Use of a rubric to evaluate student learning artifacts is a flexible and effective means to determine the effectiveness of information literacy instruction.[30] Evaluation of students' references lists is achieved through application of a rubric to student literature reviews to evaluate for quality of sources selected and integrated into the research assignment. Another, more complex evaluation technique that would offer some insights into how students located and evaluated research sources would involve the librarian and classroom instructor co-grading the students' annotated bibliographies with a rubric that evaluates students' perception of the relevance and value of a particular article or book chapter to their research question, perhaps accompanied by a short free-writing reflection exercise.

CONCLUSION

Using the lens of constructivism and its variants is an effective way to explore and interpret how students strive to become more information literate while engaging in responsible conduct of research in an interdisciplinary course. The construction and delivery of the modules of the course, Weird Science, helped students approach the intersection of research ethics and information literacy, and their learning and skills were evaluated by reviewing written work, class discussions, and student feedback. Both information literacy and responsible conduct of research are topics in service to four of the six learning goals of the Weird Science course: cultural factors that affect these disciplines; philosophical, historical, and ethical perspectives; methods for finding pertinent information; and critical evaluation of ideas and their sources. It is clear that both modules involve similar structural and conceptual elements and even share a course learning goal—the critical evaluation of ideas and their sources. This conclusion is supported when examining students' responses to the three questions posed at the end of the semester. A cluster analysis revealed two dominant clusters. One cluster included traditional information literacy concepts, such as information quality and finding relevant information. The second cluster was comprised of elements from both information literacy and responsible conduct of research, included such concepts as informed consent, giving credit where credit is due, and doing no harm.

Both modules emphasize frameworks or skill sets to help students not only complete their assignments successfully, but also help to facilitate generalization to other contexts. Given that one of the goals of interdisciplinary teaching and learning is to encourage students to bring a variety of perspectives to bear on a problem that cannot be solved within a single discipline, generalization to other contexts is key. Examples of these skill sets or frameworks include how to find quality information, what questions to ask when conducting animal or human subjects research, and where to go for more information, support a goal of generalization. When students are taken on ethical journeys informed by multiple disciplines, and they find that not only do these perspectives correspond with the course goals, but they also enhance interdisciplinary habits of mind through their generalization to contexts beyond the classroom.

NOTES

1. James M. Applefield, Richard Huber, and Mahnaz Moallem, "Constructivism in Theory and Practice: Towards a Better Understanding," *High School Journal* 84, no. 2 (December/January 2000/2001): 35–53.
2. Jean Piaget, *The Construction of Reality in the Child* (New York: Basic Books, 1954).
3. Jacqueline Grennon Brooks, "Constructivism as a Paradigm for Teaching and Learning," Constructivism as a Paradigm for Teaching and Learning. 2004, accessed July 15, 2015, http://www.thirteen.org/edonline/concept2class/constructivism/index.html
4. See Sharon J. Derry, "A Fish Called Peer Learning: Searching for Common Themes," in *Cognitive Perspectives on Peer Learning*, ed. Angela M. O'Donnell and Alison King (Mahwah, NJ: L. Erlbaum, 1999); André Kukla, *Social Constructivism and the Philosophy of Science* (London: Routledge, 2000).
5. Lev S. Vygotsky, *Mind in Society: The Development of Higher Psychological Processes* (Cambridge, MA: Harvard University Press, 1978).
6. Students are required to read Rebecca Skloot, *The Immortal Life of Henrietta Lacks* (New York: Crown Publishing, 2011).
7. National Academy of Sciences, *On Being a Scientist: A Guide to Responsible Conduct in Research* (Washington, DC: National Academies Press, 2009); Antony Gomes, Archita Saha, Poulami Datta, and Aparna Gomes. "Research Ethics for Young Researchers," *Indian Journal of Pharmacology* 45, no. 5 (September/October 2013): 540–541, doi:10.4103/0253-7613.117775.
8. Fiona Godlee, Jane Smith, and Harvey Marcovitch, "Wakefield's Article Linking MMR Vaccine and Autism Was Fraudulent," *BMJ* 342 (January 06, 2011): 7452, accessed July 21, 2015. doi:10.1136/bmj.c7452; Gregory A. Poland and Ray Spier, "Fear, Misinformation, and Innumerates: How the Wakefield Paper, the Press, and Advocacy Groups Damaged the Public Health," *Vaccine* 28, no. 12 (2010): 2361–2362, accessed July 21, 2015, doi:10.1016/j.vaccine.2010.02.052; Gregory A. Poland, "MMR Vaccine and Autism: Vaccine Nihilism and Postmodern Science," *Mayo Clinic Proceedings* 86, no. 9 (2011): 869–871, accessed July 21, 2015, doi:10.4016/33261.01.

9. Maimuna S. Majumder, Emily L. Cohn, Sumiko R. Mekaru, Jane E. Huston, and John S. Brownstein, "Substandard Vaccination Compliance and the 2015 Measles Outbreak" *JAMA Pediatrics* 169, no. 5 (2015): 494–495, accessed July 21, 2015, doi:10.1001/jamapediatrics.2015.0384.

10. Ruben Bolling, "Human Morality Made Simple," Tom the Dancing Bug. Go Comics, 1991, http://gocomics.typepad.com/tomthedancingbugblog/2014/11/human-morality-made-simple.html

11. Harry F. Harlow, "The Nature of Love," *American Psychologist* 13, no. 12 (1958): 673–685; Harry F. Harlow and Margaret Kuenne Harlow, "Learning to Love." *American Scientist* 54 (1966): 244–272.

12. John P. Gluck, "Harry F. Harlow and Animal Research: Reflection on the Ethical Paradox," *Ethics & Behavior* 7, no. 2 (1997): 149–161, accessed July 21, 2015, doi:10.1207/s15327019eb0702_6.

13. Noah Phillips, "University of Wisconsin to Reprise Controversial Monkey Studies," Wisconsinwatch.org, July 31, 2014, accessed July 22, 2015. http://wisconsinwatch.org/2014/07/university-of-wisconsin-to-reprise-controversial-monkey-studies/

14. James H. Jones, *Bad Blood: The Tuskegee Syphilis Experiment* (New York: Free Press, 1993).

15. United States Nuremberg Military Tribunals, *Trials of War Criminals Before the Nuernberg Military Tribunals Under Control Council Law* 10.2 (Washington, DC: US Government Printing Office, 1946–1949).

16. Stanley Milgram, "Behavioral Study of Obedience," *The Journal of Abnormal and Social Psychology* 67, no. 4 (1963): 371–378, doi:10.1037/h0040525; Stanley Milgram, *Obedience to Authority: An Experimental View* (New York: Harper & Row, 1974).

17. Milgram, "Behavioral Study of Obedience," 374.

18. Stanley Milgram, "Issues in the Study of Obedience: A Reply to Baumrind," *American Psychologist* 19.11 (1964): 848–852, doi:10.1037/h0044954.

19. Jerry M. Burger. "Replicating Milgram: Would People Still Obey Today?" *American Psychologist* 64.1 (2009): 1–11. doi:10.1037/a0010932; Mel Slater, Angus Antley, Adam Davison, David Swapp, Christoph Guger, Chris Barker, Nancy Pistrang, and Maria V. Sanchez-Vives, "A Virtual Reprise of the Stanley Milgram Obedience Experiments," *PLoS ONE* 1, no. 1 (December 20, 2006), accessed July 22, 2015, doi:10.1371/journal.pone.0000039.

20. Craig Haney, Curtis Banks, and Philip Zimbardo, "Interpersonal Dynamics in a Simulated Prison," *International Journal of Criminology and Penology* 1 (1973): 69–97.

21. Craig Haney, Curtis Banks, and Philip Zimbardo, "Interpersonal Dynamics in a Simulated Prison," 72.

22. City Tech librarians anticipated that the ACRL Information Literacy Framework would present a more nuanced view of the concept of authority, authorship, and expertise of creator.
23. ACRL Board, "Framework for Information Literacy for Higher Education," Framework for Information Literacy for Higher Education, February 2, 2015, accessed June 03, 2015, http://www.ala.org/acrl/standards/ilframework
24. Maura A. Smale, "Get in the Game: Developing an Information Literacy Classroom Game," *Journal of Library Innovation* 3, no. 1 (2012): 126–147.
25. Claudia J. Dold, "Critical Information Literacy: A Model for Transdisciplinary Research in Behavioral Sciences," *The Journal of Academic Librarianship* 40, no. 2 (2014): 179–184, doi:10.1016/j.acalib.2014.03.002.
26. ACRL Board, "Framework for Information Literacy for Higher Education."
27. Ibid.
28. Juris Dilevko and Lana Soglasnova, "Knowledge Maps and the Work of Academic Librarians in an Interdisciplinary Environment," *The Reference Librarian* 54, no. 2 (2013): 143–156.
29. Elizabeth J. Natalle and Kathryn M. Crowe, "Information Literacy and Communication Research: A Case Study on Interdisciplinary Assessment," *Communication Education* 62, no. 1 (2013): 97–104.
30. Ibid.

Bibliography

ACRL Board. "Framework for Information Literacy for Higher Education." *Framework for Information Literacy for Higher Education.* February 02, 2015. Accessed June 03, 2015. http://www.ala.org/acrl/standards/ilframework

Applefield, James M., Richard Huber, and Mahnaz Moallem. December/January 2000/2001. "Constructivism in Theory and Practice: Towards a Better Understanding." *High School Journal* 84 (2): 35–53.

Banuazizi, Ali, and Siamak Movahedi. 1975. "Interpersonal Dynamics in a Simulated Prison: A Methodological Analysis." *American Psychologist* 30 (2): 152–160. doi: 10.1037/h0076835.

Bolling, Ruben. 1991. "Human Morality Made Simple." *Tom the Dancing Bug. Go Comics.* http://gocomics.typepad.com/tomthedancingbug-blog/2014/11/human-morality-made-simple.html

Brooks, Jacqueline Grennon. 2004. "Constructivism as a Paradigm for Teaching and Learning." Accessed July 15, 2015. http://www.thirteen.org/edonline/concept2class/constructivism/index.html

Burger, Jerry M. 2009. "Replicating Milgram: Would People Still Obey Today?" *American Psychologist* 64 (1): 1–11. doi: 10.1037/a0010932.

Derry, Sharon J. 1999. "A Fish Called Peer Learning: Searching for Common Themes." In *Cognitive Perspectives on Peer Learning*, edited by Angela M. O'Donnell and Alison King, 197–211. Mahwah, NJ: L. Erlbaum.

Detlor, Brian, Lorne Booker, Alexander Serenko, and Heidi Julien. 2012. "Student Perceptions of Information Literacy Instruction: The Importance of Active Learning." *Education for Information* 29 (2): 147–161.

Dilevko, Juris, and Lana Soglasnova. 2013. "Knowledge Maps and the Work of Academic Librarians in an Interdisciplinary Environment." *The Reference Librarian* 54 (2): 143–156. Accessed June 05, 2015. Education Source.

Dold, Claudia J. 2014. "Critical Information Literacy: A Model for Transdisciplinary Research in Behavioral Sciences." *The Journal of Academic Librarianship* 40 (2): 179–184. doi: 10.1016/j.acalib.2014.03.002.

Gibson, Craig. 2012. "Teaching Research Across Disciplines: Interdisciplinarity and Information Literacy." In *Interdisciplinarity and Academic Libraries*, edited by Daniel C. Mack, 167–181. Chicago, IL: Association of College & Research Libraries.

Gluck, John P. 1997. "Harry F. Harlow and Animal Research: Reflection on the Ethical Paradox." *Ethics & Behavior* 7 (2): 149–161. Accessed July 21, 2015. doi: 10.1207/s15327019eb0702_6.

Godlee, F., J. Smith, and H. Marcovitch. January 06, 2011. "Wakefield's Article Linking MMR Vaccine and Autism Was Fraudulent." *BMJ* 342: 7452. Accessed July 21, 2015. doi: 10.1136/bmj.c7452.

Gomes, Antony, Archita Saha, Poulami Datta, and Aparna Gomes. September/October 2013. "Research Ethics for Young Researchers." *Indian Journal of Pharmacology* 45 (5): 540–541. doi: 10.4103/0253-7613.117775.

Haney, Craig, Curtis Banks, and Philip Zimbardo. 1973. "Interpersonal Dynamics in a Simulated Prison." *International Journal of Criminology and Penology* 1: 69–97.

Harlow, Harry F. 1958. "The Nature of Love." *American Psychologist* 13 (12): 673–685.

Harlow, Harry F., and Margaret Kuenne Harlow. 1966. "Learning to Love." *American Scientist* 54: 244–272.

Jones, James H. 1993. *Bad Blood: The Tuskegee Syphilis Experiment*. New York: Free Press.

Kuglitsch, Rebecca Z. 2015. "Teaching for Transfer: Reconciling the Framework with Disciplinary Information Literacy." *Portal: Libraries and the Academy Portal* 15 (3): 457–470. Accessed July 21, 2015. doi: 10.1353/pla.2015.0040.

Kukla, André. 2000. *Social Constructivism and the Philosophy of Science*. London: Routledge.

Majumder, Maimuna S., Emily L. Cohn, Sumiko R. Mekaru, Jane E. Huston, and John S. Brownstein. 2015. "Substandard Vaccination Compliance and the 2015 Measles Outbreak." *JAMA Pediatrics* 169 (5): 494–495. Accessed July 21, 2015. doi: 10.1001/jamapediatrics.2015.0384.

Milgram, Stanley. 1963. "Behavioral Study of Obedience." *The Journal of Abnormal and Social Psychology* 67 (4): 371–378. doi: 10.1037/h0040525.

———. 1964. "Issues in the Study of Obedience: A Reply to Baumrind." *American Psychologist* 19 (11): 848–852. doi: 10.1037/h0044954.

———. 1974. *Obedience to Authority: An Experimental View*. New York: Harper & Row.

Natalle, Elizabeth J., and Kathryn M. Crowe. 2013. "Information Literacy and Communication Research: A Case Study on Interdisciplinary Assessment." *Communication Education* 62 (1): 97–104.

National Academy of Sciences. 2009. *On Being a Scientist: A Guide to Responsible Conduct in Research*. Washington, DC: National Academies Press.

O'Connor, Lisa, and Jill Newby. 2011. "Entering Unfamiliar Territory: Building an Information Literacy Course for Graduate Students in Interdisciplinary Areas." *Reference & User Services Quarterly* 50 (3): 224–229. Accessed June 5, 2015. doi: 10.5860/rusq.50n3.224.

Phillips, Noah. "University of Wisconsin to Reprise Controversial Monkey Studies." Wisconsinwatch.org. July 31, 2014. Accessed July 22, 2015. http://wisconsinwatch.org/2014/07/university-of-wisconsin-to-reprise-contr oversial-monkey-studies/

Piaget, Jean. 1954. *The Construction of Reality in the Child*. New York: Basic Books.

Poland, Gregory A. 2011. "MMR Vaccine and Autism: Vaccine Nihilism and Postmodern Science." *Mayo Clinic Proceedings* 86 (9): 869–871. Accessed July 21, 2015. doi: 10.4016/33261.01.

Poland, Gregory A., and Ray Spier. 2010. "Fear, Misinformation, and Innumerates: How the Wakefield Paper, the Press, and Advocacy Groups Damaged the Public Health." *Vaccine* 28 (12): 2361–2362. Accessed July 21, 2015. doi: 10.1016/j.vaccine.2010.02.052.

Pritchard, Alan, and John Woollard. 2010. *Psychology for the Classroom: Constructivism and Social Learning*. London: Routledge.

Skloot, Rebecca. 2011. *The Immortal Life of Henrietta Lacks*. New York: Crown Publishing.

Slater, Mel, Angus Antley, Adam Davison, David Swapp, Christoph Guger, Chris Barker, Nancy Pistrang, and Maria V. Sanchez-Vives. December 20, 2006. "A Virtual Reprise of the Stanley Milgram Obedience Experiments." *PLoS ONE* 1 (1):1–10. Accessed July 22, 2015. doi: 10.1371/journal.pone.0000039.

Smale, Maura A. 2012. "Get in the Game: Developing an Information Literacy Classroom Game." *Journal of Library Innovation* 3 (1): 126–147. Accessed July 9, 2015. http://www.libraryinnovation.org/article/view/182/319

United States Department of Health, Education, and Welfare. 1978. "The National Commission for the Protection of Human Subjects of Biomedical and Behavioral Research." *Ethical Principles and Guidelines for the Protection of Human Subjects of Research*. Accessed July 22, 2015. http://www.hhs.gov/ohrp/humansubjects/guidance/belmont.html

United States Nuremberg Military Tribunals. 1946–1949. *Trials of War Criminals Before the Nuernberg Military Tribunals Under Control Council Law No. 10*, vol. 2. Washington, DC: US Government Printing Office.

Vygotsky, Lev S. 1978. *Mind in Society: The Development of Higher Psychological Processes*. Cambridge, MA: Harvard University Press.

CHAPTER 5

Making Connections: Writing Stories and Writing Code

Reneta D. Lansiquot and Candido Cabo

Abstract Interdisciplinary competence should be an integral part of undergraduate education. The development of interdisciplinary skills expands students' perspectives and blurs differences between general education and major courses, preparing them to be better problem-solvers in an increasingly complex and connected world. This chapter describes the design, development, and teaching of an interdisciplinary course linking creative writing and computational thinking for non-computer majors. In this interdisciplinary course, students develop original stories which they then implement as a video game prototype using computer programming. Via interdisciplinary connections between writing stories and writing computer code, even non-computer majors acquire computational thinking concepts and skills.

Keywords Computational thinking • Creative writing • Educational gamification • Integrative learning • Problem-solving • Technical writing

---

R.D. Lansiquot (✉)
Department of English, New York City College of Technology,
City University of New York, Brooklyn, NY, USA

C. Cabo
Department of Computer Systems Technology, New York City College of
Technology, City University of New York, Brooklyn, NY, USA

© The Editor(s) (if applicable) and The Author(s) 2016
R.D. Lansiquot (ed.), *Interdisciplinary Pedagogy for STEM*,
DOI 10.1057/978-1-137-56745-1_5

85

Interdisciplinary collaborations foster imaginative connections and help learners from various disciplines to enhance their creativity by harnessing the power of technology. This chapter describes interdisciplinary peda-gogical strategies used to help undergraduate students make connections between seemingly exclusive domains in a general education course, while explaining the underlying theories that informed the structure of this course.[1] Applying constructivist theory to interdisciplinary studies pro-motes learning in complex situations that allow learners to experience real-world situations and reflect on their experiences.[2] This situated, meaningful learning affords social negotiation in a community of practice as it encour-ages distributed cognition and ownership of learning as well as facilitating awareness of the knowledge construction process.[3] Research has demon-strated that interdisciplinary studies promote student learning when they purposefully connect previously disparate forms of knowledge and skills to solve problems.[4] They force students to synthesize and transfer knowl-edge across disciplinary boundaries; to comprehend the factors inherent in complex problems; to gain comfort with complexity and uncertainty; to think critically, communicate effectively, and work collaboratively; to recognize varied perspectives; and to become flexible thinkers.[5]

Cognitive flexibility theory is a critical component of constructivism as it offers a comprehensive framework for technology-based learning in complex domains, such as the interconnection between creative writing and computer programming, while avoiding oversimplification.[6] As an application of this theory, the next section of this chapter presents our research methods and findings that led, over a five-year period, to the creation of a co-taught, writing-intensive interdisciplinary liberal arts and science course at the general education level.

Problem-Solving with Computer Programming

We initiated this educational research study because first-year problem-solving and computer programming gateway courses had low pass rates, causing computer science majors to drop out or transfer to other majors. The urban institution where this research took place, New York City College of Technology (City Tech), one of the senior colleges of the City University of New York, serves mostly underrepresented minority students who have little or no previous computer programming experience and weak mathematical backgrounds. City Tech offers a computer problem-solving course (PS) to prepare students in computer science majors for the more rigorous first programming course (CS1).[7] After the department changed the educational technology that it used for computer programming, from a text-based proprietary software to the publicly available *Alice* application (www.alice.org, a computer programming environment that supports the creation of three-dimensional animations as a video game prototype), the pass rate increased by almost 10 %.[8] More significantly, this higher pass rate in the *Alice* PS course was not accompanied by any decline in students' readiness for the subsequent CS1 course.[9]

In conjunction with these research studies, we created an interdisciplinary learning community[10] (LC) and used it as an educational research incubator. As part of a LC linked to a first course in English composition with strong narrative components, the *Alice* PS course further increased student performance and retention,[11] as evidenced by computer science majors' achievement in their introductory computer courses. Interestingly, English composition is also a gateway course with low passing rates for computer science majors: their performance in English is lower than their performance in the PS course. Linking the English composition course with a computer programming course in a LC resulted in equal amounts of improvement in both English composition and the computer programming course.[12] In the LC, students use the problem-solving, programming, and writing abilities gained in the linked courses to produce a video game prototype.[13] This interdisciplinary pedagogical approach allows students to connect knowledge and skills from across disciplines, developing synergies between writing stories and writing computer programs. Writing stories and implementing those stories as a computer program to produce a video game prototype provides an engaging learning context for our students.

COMPUTATIONAL THINKING FOR ALL STUDENTS

Based on the pervasiveness of computing in today's world, it could be argued that computational thinking (i.e., the application of computing concepts and skills to solve problems, not only in computer science but also in other disciplines) should be a part of a twenty-first century liberal education for a broad range of college students, including those not majoring in computing.[14] Computational thinking can help students to frame problems in a variety of fields and disciplines (not just STEM disciplines) and, in so doing, to become better problem-solvers in their professions.

Currently, many students not majoring in computer science at City Tech take the first-year PS course described above to satisfy the computer literacy requirement in their major or to learn computational thinking concepts. However, since PS is a gateway course for computer science majors, it is even more challenging for non-majors. In a recent assessment of computer programming concepts and skills, 44 % of computer science majors taking the PS course (without being a part of a LC) demonstrated an adequate understanding of computer programming concepts and could write viable programs. When computer majors take PS as part of an LC, the portion of students with adequate performance increases to 56 %.[15] In contrast, only 30 % of non-computer majors taking the PS course (PS courses for non-computer majors are not part of a LC) perform adequately in computer programming concepts and skills. The results of the assessments above indicate that teaching computational thinking concepts and skills to non-computer majors requires pedagogical strategies which are different than those that may work with computer science majors.

Two major issues are involved in designing a computational thinking course (or a course introducing computational thinking elements) for a broad range of college students, including non-computer majors: (1) what concepts and skills to include in the course and (2) what learning context and pedagogical approach to use in order to make computational thinking more accessible.[16] In our view, a computational thinking course should include a combination of procedural and object-oriented programming concepts, including the steps required in using computers to solve a problem and the use of flowcharting techniques and such programming structures as sequencing, repetition loops, and decision statements to solve an algorithm. It is also important that students are introduced to concepts of object-oriented programming like classes, objects, properties, and methods. The selection of the learning context and pedagogical approach used to teach those concepts results from our experience in linking writing and computer programming in the interdisciplinary LC.[17] Just as an interdisciplinary context linking writing and computer programming was beneficial for computer majors, it can also contribute to facilitate the learning of computational thinking concepts and skills for non-computer majors.

PROGRAMMING NARRATIVES: COMPUTER-ANIMATED STORYTELLING

To introduce computational and interdisciplinary thinking for non-computer majors, we created an interdisciplinary, writing-intensive course called Programming Narratives: Computer-Animated Storytelling, in which students integrate problem-solving, writing, and computational thinking to produce a narrative-driven video game prototype. In this course, students write original video game stories and then present their ideas to their classmates. Upon successful completion of the course, students should be able to create story concept maps[18]; demonstrate an understanding of the structure of game stories; exhibit understanding of the steps required in solving a problem using a computer; demonstrate understanding of flowcharting techniques to solve an algorithm; program using sequencing, repetition loops, and decision statements; demonstrate an understanding of object-oriented programming; use a range of language (from formal to informal) as appropriate for the subject, purpose, and audience; demonstrate understanding of various narrative structures; write, revise, and proofread clear and logical sentences using correct spelling, conventional punctuation, and correct grammar and syntax; use varied sentence structure; order and connect sentences and paragraphs effectively, using transitions and parallelism; and cite sources within the text and on a reference page using appropriate documentation format.

In order to explore the concept of "story," students are introduced to Aristotle's *Poetics* and his six elements of drama (plot, theme, character, diction/language/dialogue, music/rhythm, and spectacle), "unity of action" (i.e., exposition, rising action, climax, falling action, and resolution), stages of the plot (complication and unraveling), and types of conflict in various media genres, emphasizing those found in recent movies and classic stories. Students then learn Joseph Campbell's theory of the hero's journey, a structure as common in ancient epics such as *Gilgamesh*, *Beowulf*, *The Iliad*, and *The Odyssey* as in more modern works like the *Star Wars* series and the *Harry Potter* books and movies.[19] This narrative structure works especially well for heroic quests, epic adventures, and journeys of enlightenment.[20] The hero's journey contains three acts: the departure, the initiation, and the return. These acts can be further broken down into several stages, which Campbell depicts as a counterclockwise movement from the Ordinary World to a series of essential stages: Call to Adventure; the Refusal of the Call; Meeting the Mentor; Crossing the Threshold; Tests, Allies, and Enemies; Approaching the Inmost Cave; The Ordeal; Reward; The Road Back; Resurrection; Return with the Elixir; and Return to the Ordinary World.[21] Students choose a few story ideas to develop and then revise these stories collaboratively, initially using concept maps to represent the current story. Finally, each student develops an engaging character side-quest (i.e., a branching story path) and an accompanying concept map, using the *Visual Understanding Environment* (vue.tufts.

edu, a publicly available application). Students must also present a ratio-nale explaining why the side-quest is meaningful for the protagonist as well as for the target audience of the game.

Short narrative readings of various kinds are assigned to help students make connections between classic literature and modern styles of story-telling—that is, between general education and computer science. These works range from Leo Tolstoy's "The Three Questions" and Sophocles' *Oedipus the King* to Richard Connell's "The Most Dangerous Game" and Ray Bradbury's "A Sound of Thunder."

The structure of narratives and concepts of problem-solving are intro-duced by using the logical constructs inherent in computer programming languages. Students collaboratively implement their stories using *Alice* and they learn computational thinking concepts along the way. The game design tasks require both creative writing and computational thinking skills. The concepts and skills introduced in this interdisciplinary course are thus embedded in a complex, meaningful project that students choose and are also connected to concepts and skills developed in the creative and technical writing process.

We chose to have students develop their own video games (with nar-rative, action-adventure, and role-playing components) rather than using off-the-shelf educational games because, through the game development process, students engage in meaningful learning while observing how their own written text actually functions in situated, embodied, active, and critical ways as the game players read it.[22] Also, avoiding the use of existing educational games that students may have already experienced reduced the risk of preconceived negative reactions to the games that could stifle their critical examination and exploration. Students engage in both cre-ating games and playing their classmates' game prototypes. To promote engaged learning, students meet several project milestones:

1. Prepare a flowchart and concept map of the video game prototype;
2. Write and program a setting for the video game;
3. Write and program characters (protagonist and antagonist) for the video game;
4. Integrate video game setting and characters;
5. Implement the characters' interactions among themselves and with their world (the story);
6. Develop individually a character side-quest within the group-developed video game;

7. Use events to allow user interactivity with the story (the game); and
8. Integrate the main story and side-quest.

This combination of clear goals, challenging tasks, explicitly stated performance standards compels students to collaborate with each other and to generate original, creative stories.[23]

ENGAGED LEARNING AND GAME DESIGN

This challenging set of goals made developing performance assessments somewhat challenging as well. The final game design document, which accompanies student groups' video game prototype or trailer as part of the final project, includes sections on the analysis conducted (video game narrative, target audience, delivery platform, and review of competing games), design (player characteristics, game mechanics, and challenge), and project description (video game prototype, review of relevant literature, pseudocode, concept maps, and storyboards). This set of tasks gives students ample opportunity to demonstrate that they have fulfilled the learning outcomes. First, in the analysis section, students summarize the group's revised version of the entire game story, including all individual group member side-quests. They answer questions such as who the target audience is, why the player should care emotionally about the protagonist, and why the story is socially relevant or engaging for the proposed target audience. Students also describe their reasoning for the delivery platform selected (e.g., a game console such as Sony's PlayStation or Microsoft's Xbox, a desktop PC or Mac, or a mobile app), and review competing games. In the design section, students describe player characteristics, game mechanics, challenges, possible setting, and hooks—that is, the features that cause people to keep playing, which can be characterized as action, time, resource, tactical, and strategic.

Finally, in the project description section, students explain the scenes included in their video game trailer and why they were selected. They review literature from the course and provide their own answer to the question of what makes a good story. As a group, students summarize, analyze, and synthesize the short stories discussed during the semester, using the hero's journey and literary devices as interpretive tools. They also comment on what makes a good story and relate their answer to the features of their own video game narrative. Students are asked to persuade their readers that the game's narrative will be engaging to their target audience, as well as to articulate how they would promote the game to the target audience. This final section of the game design document also includes pseudocode, concept maps, and storyboards (in the form of screenshots from *Alice*).

A game design document is a powerful pedagogical tool to facilitate interdisciplinary studies. Powerful learning experiences occur as the students compare their own newly developed narratives to those of classic literature. By way of illustration, one student group enrolled in this course wrote and coded elements of a role-playing game, *Conspiracy*, which is set in their hometown and summarized below:

> A conspiracy lies behind the structures of power on Earth. Unbeknownst to the majority of the world's population, all the world's leaders in politics and art are members of a Reptilian army sent to Earth to lull the native population into servitude. The hero of our game, Liz, is a high school intern working at New York's City Hall who accidentally uncovers the Reptilians' secret. Guided by YouTube conspiracy theorists, Liz travels to secret New York City locations to unlock the clues to the Reptilian plan for world domination. Will she solve the riddles and thwart the aliens' plan before the mother ship arrives to carry out their final plan? It's up to you.

In the literature review of their game design document, this group wrote that the common threads running through classic stories of the hero's journey include "the elements of suspense, foreshadowing, and imagery. The students then analyzed the use of these elements in seven works: "The Three Questions" by Leo Tolstoy, *Oedipus the King* by Sophocles, "The Lottery" by Shirley Jackson, "The Most Dangerous Game" by Richard Connell, "The Lady or the Tiger" by Frank Stockton, "Young Goodman Brown" by Nathaniel Hawthorne, and "Sound of Thunder" by Ray Bradbury. In the case of the last three of these works,[24] they noted that foreshadowing is used to help the reader envision "possible conclusions" for these short stories. The students then mapped the hero's journey to *Oedipus the King* and Conspiracy:

> In our story, Conspiracy, there is the element of suspense since the reader wants to find out what will happen next to Liz. The reader wants to find out about her discoveries as well as who ultimately triumphs in the end. Foreshadowing is also used throughout the story. For example, the sushi that Carol always eats, the hissing of the coffee machines, the green streak in Tillie's hair and many more. The descriptive details as well as the con-

stant action in the story serve to draw readers in and make them want to read further. Conspiracy also adheres to the hero's journey. The Ordinary World is Liz in her internship. The Call to Adventure occurs when Liz walks into the meeting and finds the Reptilian people. There is no Refusal of the Call. Meeting the Mentor is when Liz finds David Ickes, her online mentor, who gives her more information about the Reptilian people. Crossing the Threshold is when Liz sets out to find out whatever she can about the Reptilians. Tests, Allies and Enemies are the events in the Lair, the chase over the bridge, the fight with Tillie, and the challenge of saving the baby from the Reptilians. Approaching the Innermost Cave is when Liz finds herself in the Barclay Center. The Ordeal is when Liz has to fight off the Queen of the Reptilians. The reward is that the Queen is dead and everyone is safe. The Road Back is when the Reptilians flee. The Resurrection is that the world is now safe and David Ickes becomes a leader of a new world government that allows the people of the world to coexist in harmony. Return with the Elixir is when Liz graduates and learns the true identity of her guidance counselor.

The students concluded, "With the use of suspense and foreshadowing, Conspiracy will engage the audience and pull them into the story. Conspiracy also follows the hero's journey, which is used in good stories. Therefore, we believe that Conspiracy is a good story that many people will find interesting and provides a good backstory for our computer game."

This group, like most others, decided to show several of the exciting scenes of their game prototype. Their trailer began by introducing the main character and setting up the basic premise of the story (i.e., Liz's discovery of the reptilian plot). They explained, "The trailer contains several action scenes of Liz battling the reptilian horde on the Brooklyn Bridge, at the Brooklyn Museum, and in the final confrontation with the main antagonist, the reptilian queen. ... [It] prominently features locations in New York City, giving the game an appeal of verisimilitude and showing how the game incorporates real locations and people into its narrative and game play." Since the video game's progression is based on a computer program, to do all this the students had to make meaningful connections between creative writing and computational thinking.

As another student group wrote and coded portions of a sci-fi action-adventure game with stealth elements, called Life of Lineage. Following is a portion of their game summary:

The world of Bellum is very much like our own, except that the inhabitants are naturally able to use paranormal abilities by adolescence. Through constant war, an emerging one-world government, Mala Fide, uses advanced nanotechnology to suppress these abilities and begins its reign of order. A resistance called "The Vox" has formed against the government. Jack, Seraph, and Genki are a part of this resistance, and they are currently on a rescue mission to rescue an undercover agent. Fighting for the right of personal autonomy and individual freedom, this group has devised hacking techniques to circumvent the suppression by Mala Fide. Their goal is to extract their undercover agent, Blink, before she spills vital intelligence to the enemy. Their covert mission, infiltrating a heavily guarded medical research facility, seems impossible.

In the literature review of their game design document, this group stated, "A good story also requires a player to develop an emotional connection with the characters," and "attention to details is also a primordial aspect of any good story." This attention to detail is fostered as the students analyze and develop stories through appropriate literacy devices, writing, and coding.[25]

The trailer for this game showed panoramas of the game world and included detailed character movements that were extremely time-consuming to code. For instance, this group coded close-up facial gestures as well as a scene that required the character to enter an access code in order to unlock a door. Their trailer also included recorded voice-acting and sound effects. Students pushed the limits of what can be easily accomplished using the free educational tool *Alice* by creating an illusion of an elevator that the character can enter and control.

By the time that students presented their video game trailers, they were invested in their creation, as reflected in frequent statements that they did not want to "give away" specific aspects of their game in the trailer. Although the complete Life of Lineage game would likely never be completed (at least not before the course ended), during the group's trailer presentation one member—who was planning to become a legal assistant, not a technical wizard—said that she wanted her classmates to play it. During class discussions, other students asked the presenters how they accomplished certain effects, displaying interest in the coding skills that undergirded the animations. For example, one student explained that, to show spilled blood, he created "a red circle that moves up from the floor and expands." Students put the theory of distributed cognition, which focuses on the social and physical settings of learning, into practice.[26] This theory's incorporation of both the social and physical worlds makes it very useful in analyzing human-computer interactions and educational technologies. Overall, student groups functioned like a production studio, with each group member contributing his or her skills to game development.

General Education, Computational Thinking, and Interdisciplinary Awareness

Students at our institution must take a combination of courses in their major/discipline comprising about two-thirds of the course credits required for graduation. General education courses in the liberal arts and sciences comprise the remaining one-third to meet graduation requirements. Each degree program has well-defined learning outcomes mapped to the learning outcomes of each course within the program. Learning outcomes usually expand across individual courses. Students must consider the different courses required for graduation as part of a single whole, the degree program. They then can establish synergistic connections between the different courses of the curriculum.

In reality, students often do not relate courses in their major to general education courses. Consequently, they cannot transfer skills between them. This lack of transfer is not a problem unique to our institution. Abundant evidence suggests that the transfer of skills between courses is relatively rare.[27] The problem occurs not only between general education and the major, but also between courses in the major. Many factors hinder this transfer. Students may have forgotten some of the material learned in a previous course, they may not perceive the connections, they may see the connections but cannot use the material in meaningful ways in a different context, or the instructor's pedagogical approach may not foster transfer.[28]

Our computational thinking writing-intensive interdisciplinary course helps students break disciplinary barriers by making connections among the domains of classic literature, creative writing, computer programming, and technical writing, while meeting our institutional general education learning outcomes in the categories of "skills," "integration," and "values, ethics, relationships." In the skills category, this course fosters students' communication skills, their ability to communicate in diverse settings and groups using written (both reading and writing), oral (both speaking and listening), and visual means, and in more than one language. The course also fosters students' inquiry/analysis skills as well as their ability to employ scientific reasoning and logical thinking. In the integration category, this interdisciplinary course allows students to work productively within and across disciplines; that is, they can make meaningful and multiple connections among the liberal arts and between the liberal arts and the areas of study leading to a major or profession. This course also promotes students' information literacies as they gather, interpret, evaluate, and apply information discerningly from a variety of sources. Finally, in the values, ethics, relationships category, this course promotes professional and personal development as students work with teams, including those of diverse composition, building consensus while respecting and using creativity.

The first iteration of this co-taught, interdisciplinary course on writing and computational thinking enrolled a total of 40 students from a wide variety of majors—architectural technology, communication design, entertainment technology, hospitality management, and mechanical engineering, to name a few. Preliminary findings from the course suggest that the learning objectives and the pedagogical approach used seem adequate for a broad range of non-computer majors.[29] Performance on writing and computing assessments as well as final grades (88 % of students obtained a grade of C or better) indicated that a vast majority of students successfully achieved the learning objectives. These results were consistent with student perceptions as reflected in an end-of-course survey. About 80 % of students agreed or strongly agreed with each of the following statements: "I understand the various narrative structures"; "I understand the structure of video game stories"; "I understand the steps required to solve a problem with a computer"; "I understand concepts of object-oriented programming"; and "I can program using sequencing, selection, and repetition structures."

Students satisfactorily integrated creative writing and computer programming to develop their video game prototypes, making in-depth interdisciplinary connections along the way. Specifically, 75 % of students responded affirmatively to the question, "Were you able to make interdisciplinary connections between writing stories and writing code?" Further analysis of survey responses revealed that students who reported playing video games regularly in their personal life were more readily able to make these connections. Of the subgroup (53 % of all students) who played video games regularly, 88 % indicated that they were able to find interdisciplinary connections, in contrast to only 60 % of those who did not play video games. Of the 25 % who did not find interdisciplinary connections, three-fourths were not gamers. The relationship between playing video games and interdisciplinary awareness is intriguing and merits further study.

CONCLUSION

We believe that interdisciplinary competence should be an integral part of undergraduate education. Interdisciplinary courses help students make connections between courses in general education and their majors, and between courses in their majors. Here, we describe our experience with the design, development, and teaching of an interdisciplinary course linking creative writing and computational thinking. In the course, students develop original stories which they later implement as a video game prototype using computer programming. This interdisciplinary approach seems to be effective in teaching computational thinking concepts and skills to non-computer majors. Moreover, students were able to make interdisciplinary connections between creative writing and computational thinking.

NOTES

1. See David R. Krathwohl, "A Revision of Bloom's Taxonomy: An Overview," *Theory into Practice* 41, no. 4 (2002): 212–218. In this revision of Bloom's taxonomy, creating is paramount.

2. Jean Piaget, *To Understand Is To Invent: The Future of Education* (New York: Grossman, 1973).

3. See Jean Lave and Etienne Wenger, *Situated Learning: Legitimate Peripheral Participation* (Cambridge: Cambridge University Press, 1991); Brown, Ann L., Doris Ash, Martha Rutherford, Kathryn Nakagawa, Ann Gordon, and Joseph C. Campione. "Distributed Expertise in the Classroom," in *Distributed Cognitions: Psychological and Educational Considerations*, ed. Gavriel Salomon (Cambridge: Cambridge University Press, 1997), 188–228.

4. Project Kaleidoscope, *What Works in Facilitating Interdisciplinary Learning in Science and Mathematics: Summary Report* (Washington, DC: AAC&U, 2011); Lisa R. Lattuca, Lois J. Voigt, and Kimberly Q. Fath, "Does Interdisciplinarity Promote Learning? Theoretical Support and Researchable Questions," *Review of Higher Education* 28, no. 1 (2004): 23–48.

5. See Susan Elrod and Mary J. S. Roth, *Leadership for Interdisciplinary Learning: A Practical Guide to Mobilizing, Implementing, and Sustaining Campus Efforts* (Washington, DC: Association of American Colleges and Universities, 2012); Project Kaleidoscope, *What Works in Facilitating Interdisciplinary Learning in Science and Mathematics: Summary Report*; Lisa R. Lattuca, *Creating Interdisciplinarity: Interdisciplinary Research and Teaching among College and University Faculty* (Nashville, TN: Vanderbilt University Press, 2001); Reneta D. Lansiquot, Candido Cabo, and Tamrah D. Cunningham, "Playing between the Lines: Promoting Interdisciplinary Studies with Virtual Worlds," in *The Role-Playing Society: Essays on the Popular Influence of Role-Playing Games,*

eds. Andrew Byers and Francesco Crocco (Jefferson, NC: McFarland, 143–162.

6. These seminal concepts are discussed in Rand J. Spiro, Paul J. Feltovich, Michael J. Jacobson, and Richard L. Coulson, "Cognitive Flexibility, Constructivism, and Hypertext: Random Access Instruction for Advanced Knowledge Acquisition in Ill-Structured Domains," in *Constructivism and the Technology of Instruction: A Conversation*, eds. Thomas M. Duffy and David H. Jonassen (Hillsdale, NJ: Lawrence Erlbaum Associates, 1992), 57–76.

7. For a detailed description of these courses, see Candido Cabo and Reneta D. Lansiquot, "Synergies between Writing Stories and Writing Programs in Problem-Solving Courses," in *2014 IEEE Frontiers in Education (FIE) Conference*. (New York: IEEE, 2014), 888–896.

8. Ibid.

9. Ibid.

10. Learning communities encourage integration of learning across courses. See George D. Kuh, *High-Impact Educational Practices: What They Are, Who Has Access to Them, and Why They Matter* (Washington, DC: AAC&U, 2008).

11. Cabo and Lansiquot, "Synergies between Writing Stories and Writing Programs in Problem-Solving Courses."

12. For further details of our mixed-methodology interdisciplinary studies, see Cabo and Lansiquot, "Synergies between Writing Stories and Writing Programs in Problem-Solving Courses"; Candido Cabo and Reneta D. Lansiquot, "Development of Interdisciplinary Problem-solving Strategies through Games and Computer Simulations," in *Cases on Interdisciplinary Research Trends in Science, Technology, Engineering, and Mathematics: Studies on Urban Classrooms*, ed. Reneta D. Lansiquot (New York: Information Science Reference, 2013), 268–294; Reneta D. Lansiquot and Candido Cabo. "Alice's Adventures in Programming Narratives," in *Transforming Virtual Learning: Cutting-Edge Technologies in Higher Education*, vol. 4, ed. Randy Hinrichs and Charles Wankel (Bingley, UK: Emerald, 2011), 311–331; Reneta D. Lansiquot and Candido Cabo, "Strengthening the Narrative of Computing with Learning Communities," in *Proceedings of the World Conference on Educational Media and Technology 2014* (Chesapeake, VA: AACE, 2014), 2217–2223; Reneta D. Lansiquot and Candido Cabo, "The Narrative of Computing," in *Proceedings of the World Conference on Educational Media and Technology 2010* (Chesapeake, VA: AACE, 2010), 3655–3660.

13. For examples of games created by students in this course, see Reneta D. Lansiquot, Ashwin Satyanarayana, and Candido Cabo, "Using Interdisciplinary Game-Based Learning to Develop Problem Solving and Writing Skills," in *Proceedings of the 121st American Society for Engineering Education Annual Conference* (Washington, DC: ASEE, 2014).

14. Mark Guzdial, "Paving the Way for Computational Thinking," in *Communications of the ACM* 51, no. 8 (2008): 25–27; Jeannette M. Wing, "Computational Thinking," in *Communications of the ACM* 49, no. 3 (2006): 33–35.

15. Cabo and Lansiquot, "Synergies between Writing Stories and Writing Programs in Problem-Solving Courses."

16. Guzdial, "Paving the Way for Computational Thinking."

17. Lansiquot and Cabo, "Strengthening the Narrative of Computing with Learning Communities."

18. This is an illustration created to represent the elements of a plot and/or the relationships between characters. See Reneta D. Lansiquot and Candido Cabo. "Concept Mapping Narratives to Promote CSCL and Interdisciplinary Studies," in *Proceedings of the Computer Supported Collaborative Learning (CSCL) Conference 2015*, vol. 2 (New York: ACM, 2015), 504–507.

19. Josiah Lebowitz and Chris Klug, *Interactive Storytelling for Video Games: A Player-Centered Approach to Creating Memorable Characters and Stories* (New York: Taylor & Francis, 2011).

20. Ibid.

21. The hero's journey is explained in Joseph Campbell, *The Hero with a Thousand Faces* (New York: Pantheon, 1949). This plot structure is explored in Christopher Vogler, *The Writer's Journey: Mythic Structures for Writers*, 3rd ed. (Studio City, CA: Michael Wiese, 2007).

22. See James Paul Gee, *What Video Games Have to Teach Us About Learning and Literacy* (New York: Palgrave Macmillan, 2003); Isabela Granic et al., "The Benefits of Playing Video Games," *American Psychologist* 69, no. 1 (2013): 66–78; Merrilea J. Mayo, "Video Games: A Route to Large-Scale STEM Education?" *Science* 323, no. 5910 (2009): 79–82.

23. See Michele D. Dickey, "Engaging by Design: How Engagement Strategies in Popular Computer and Video Games Can Inform Instructional Design," *Educational Technology Research and Development* 53, no. 2 (2005): 67–83.

24. "The Lady or the Tiger" is an open-ended story that allows the reader to decide the fate of the main characters. Although "Young Goodman

Brown" has a clear conclusion, the reader is left uncertain whether the main character imagined most of the story. "Sound of Thunder" also leaves the reader uncertain as to the consequences of time travel and the realities of the world at the end of the story.

25. This pedagogical strategy and two others are described in Reneta D. Lansiquot and Candido Cabo, "Strategies to Integrate Writing in Problem-Solving Courses: Promoting Learning Transfer in an Interdisciplinary Context," in *Proceedings of the 122nd American Society for Engineering Education Annual Conference* (Washington, DC: ASEE, 2015).

26. See Lave and Wenger, *Situated Learning*, and the zone of proximal development (ZDP) as discussed in Lev S. Vygotsky, *Mind in Society: The Development of Higher Psychological Processes* (Cambridge, MA: Harvard University Press, 1978). ZDP is the difference between what students can do alone and what students can do with assistance.

27. Susan M. Barnett and Stephen J. Ceci, "When and Where Do We Apply What We Learn? A Taxonomy for Far Transfer," *Psychological Bulletin* 128, no. 4 (2002): 612–637; Ruth Benander and Robin Lightner, "Promoting Transfer of Learning: Connecting General Education Courses," *The Journal of General Education* 54, no. 3(2005): 199–208; Erik De Corte, "Transfer as the Productive Use of Acquired Knowledge, Skills, and Motivations," *Current Directions in Psychological Science* 12, no. 4 (2003): 142–146.

28. Benander and Lightner, "Promoting Transfer of Learning: Connecting General Education Courses."

29. One section of this course was co-taught by the authors of this chapter; the other section was taught by the first author of this chapter and a graduate student in game design, Tamrah D. Cunningham, who participated in the first cohort of the aforementioned LC as an undergraduate. An account of her experiences is included in Lansiquot, Cabo, and Cunningham, "Playing between the Lines: Promoting Interdisciplinary Studies with Virtual Worlds," as well as in the concluding chapter of the present book.

BIBLIOGRAPHY

Alice [software]. Accessed July 10, 2015. http://www.alice.org

Barnett, Susan M., and Stephen J. Ceci. 2002. "When and Where Do We Apply What We Learn? A Taxonomy for Far Transfer." *Psychological Bulletin* 128 (4): 612–637.

Benander, Ruth, and Robin Lightner. 2005. "Promoting Transfer of Learning: Connecting General Education Courses." *The Journal of General Education* 54 (3): 199–208.

Bloom, Benjamin S., Max B. Englehart, Edward J. Furst, Walter H. Hill, and David R. Krathwohl. 1956. "Taxonomy of Educational Objectives, the Classification of Educational Goals." In *Handbook I: Cognitive Domain*, edited by Benjamin S. Bloom. New York: McKay.

Brown, Ann L., Doris Ash, Martha Rutherford, Kathryn Nakagawa, Ann Gordon, and Joseph C. Campione. 1997. "Distributed Expertise in the Classroom." In *Distributed Cognitions: Psychological and Educational Considerations*, edited by Gavriel Salomon, 188–228. Cambridge: Cambridge University Press.

Cabo, Candido, and Reneta D. Lansiquot. 2013. "Development of Interdisciplinary Problem-Solving Strategies through Games and Computer Simulations." In *Cases on Interdisciplinary Research Trends in Science, Technology, Engineering, and Mathematics: Studies on Urban Classrooms*, edited by Reneta D. Lansiquot, 268–294. New York: Information Science Reference.

———. 2014. "Synergies between Writing Stories and Writing Programs in Problem-Solving Courses." In *2014 IEEE Frontiers in Education (FIE) conference*, 888–896. New York: IEEE.

Campbell, Joseph. 1949. *The Hero with a Thousand Faces*. New York: Pantheon.

De Corte, Erik. 2003. "Transfer as the Productive Use of Acquired Knowledge, Skills, and Motivations." *Current Directions in Psychological Science* 12 (4): 142–146.

Dickey, Michele D. 2005. "Engaging by Design: How Engagement Strategies in Popular Computer and Video Games Can Inform Instructional Design." *Educational Technology Research and Development* 53 (2): 67–83.

Elrod, Susan, and Mary J. S. Roth. 2012. *Leadership for Interdisciplinary Learning: A Practical Guide to Mobilizing, Implementing, and Sustaining Campus Efforts*. Washington, DC: AAC&U.

Gee, James P. 2013. *What Video Games Have To Teach Us about Learning and Literacy*. New York: Palgrave.

Granic, Isabela, Adam Lobel, and Rutger C. M. E. Engels. 2013. "The Benefits of Playing Video Games." *American Psychologist* 69 (1): 66–78.

Guzdial, Mark. 2008. "Paving the Way for Computational Thinking." *Communications of the ACM* 51 (8): 25–27.

Krathwohl, David R. 2002. "A Revision of Bloom's Taxonomy: An Overview." *Theory into Practice* 41 (4): 212–218.

Kuh, George D. 2008. *High-Impact Educational Practices: What They Are, Who Has Access to Them, and Why They Matter*. Washington, DC: AAC&U.

Lansiquot, Reneta D., and Candido Cabo. 2010. "The Narrative of Computing." Proceedings of the world conference on educational media and technology 2010, 3655–3660. Chesapeake, VA: AACE.

————. 2011. "Alice's Adventures in Programming Narratives." In *Transforming Virtual Learning: Cutting-Edge Technologies in Higher Education*, edited by Randy Hinrichs and Charles Wankel, vol. 4, 311–331. Bingley, UK: Emerald.

————. 2014. "Strengthening the Narrative of Computing with Learning Communities." Proceedings of the world conference on educational media and technology 2014, 2217–2223. Chesapeake, VA: Association for the Advancement of Computing in Education.

————. 2015a. "Concept Mapping Narratives to Promote CSCL and Interdisciplinary Studies." Proceedings of the Computer Supported Collaborative Learning (CSCL) conference 2015, vol. 2, 504–507. New York: ACM.

————. 2015b. "Strategies to Integrate Writing in Problem-Solving Courses: Promoting Learning Transfer in an Interdisciplinary Context." Proceedings of the 122nd American Society for Engineering education annual conference. Washington, DC: ASEE.

Lansiquot, Reneta D., Ashwin Satyanarayana, and Candido Cabo. 2014. "Using Interdisciplinary Game-Based Learning to Develop Problem Solving and Writing Skills." Proceedings of the 121st American Society for Engineering education annual conference. Washington, DC: ASEE.

Lansiquot, Reneta D., Candido Cabo, and Tamrah D. Cunningham. 2016. "Playing between the Lines: Promoting Interdisciplinary Studies with Virtual Worlds." In *The Role-Playing Society: Essays on the Popular Influence of Role-Playing Games*, edited by Andrew Byers and Francesco Crocco, 143–162. Jefferson, NC:McFarland.

Lattuca, Lisa R. 2001. *Creating Interdisciplinarity: Interdisciplinary Research and Teaching among College and University Faculty*. Nashville, TN: Vanderbilt University Press.

Lattuca, Lisa R., Lois J. Voigt, and Kimberly Q. Fath. 2004. "Does Interdisciplinarity Promote Learning? Theoretical Support and Researchable Questions." *Review of Higher Education* 28 (1): 23–48.

Lave, Jean, and Etienne Wenger. 1991. *Situated Learning: Legitimate Peripheral Participation*. Cambridge: Cambridge University Press.

Lebowitz, Josiah, and Chris Klug. 2011. *Interactive Storytelling for Video Games: A Player-Centered Approach to Creating Memorable Characters and Stories*. New York: Taylor & Francis.

Mayo, Merrilea J. 2009. "Video Games: A Route to Large-Scale STEM Education?" *Science* 323 (5910): 79–82.

Piaget, Jean. 1973. *To Understand Is To Invent: The Future of Education*. New York: Grossman.

Project Kaleidoscope. 2011. *What Works in Facilitating Interdisciplinary Learning in Science and Mathematics: Summary Report*. Washington, DC: AAC&U.

Spiro, Rand J., Paul J. Feltovich, Michael J. Jacobson, and Richard L. Coulson. 1992. "Cognitive Flexibility, Constructivism, and Hypertext: Random Access

Instruction for Advanced Knowledge Acquisition in Ill-Structured Domains." In *Constructivism and the Technology of Instruction: A Conversation*, edited by Thomas M. Duffy and David H. Jonassen, 57–76. Hillsdale, NJ: Lawrence Erlbaum Associates.

Visual Understanding Environment [software]. Accessed July 20, 2015. http:// vue.tufts.edu

Vogler, Christopher. 2007. *The Writer's Journey: Mythic Structures for Writers*. 3rd ed. Studio City, CA: Michael Wiese.

Vygotsky, Lev S. 1978. *Mind in Society: The Development of Higher Psychological Processes*. Cambridge, MA: Harvard University Press.

Wing, Jeannette M. 2006. "Computational Thinking." *Communications of the ACM* 49 (3): 33–35.

.

CHAPTER 6

Authenticating Interdisciplinary Learning through a Geoscience Undergraduate Research Experience

Reginald A. Blake and Janet Liou-Mark

Abstract The geosciences may be the most interdisciplinary of all STEM disciplines. Earth system sciences and the complex sub-systems of the cryosphere, the atmosphere, the lithosphere, the biosphere, and the hydrosphere subsume all human activity and are critical to every aspect of life on Earth. Therefore, by their very natures, the geosciences are rife with copious interdisciplinary strands and themes that are waiting to be explored by students from a wide range of STEM disciplines. With studies highlighting the benefits of authentic undergraduate research experiences, an innovative program was designed to have STEM students actively and collaboratively construct their knowledge of the geosciences. Results indicate that the geoscience research experiences increased the students' understanding of the relevancy of their interdisciplinary study to society.

Keywords Interdisciplinary • Geosciences • Remote sensing • STEM • Undergraduate research • Underrepresented minority students

R.A. Blake
Department of Physics, New York City College of Technology,
City University of New York, Brooklyn, NY, USA

J. Liou-Mark (jliou-mark@citytech.cuny.edu ✉)
Department of Mathematics, New York City College of Technology,
City University of New York, Brooklyn, NY, USA

Globalization and the need to address the diverse complexities and maladies of an ever-advancing yet ever-shrinking world are the main drivers behind the imperative for interdisciplinary learning at this critical juncture of the twenty-first century. Students now need not only to think, navigate, and advance on the local scale, but also to be equipped with the knowledge and the skills necessary to confront global challenges. These challenges in science, technology, engineering, and mathematics (STEM), medicine, economics, and a whole host of other disciplines are no longer restricted to and confined within rigid disciplinary borders. Indeed, historical disciplinary borders are now not only blurred, but in some cases they have also been torn down. Disciplines (and knowledge for that matter) are no longer isolated and siloed; they are systematic, integrative, and interdisciplinary in nature and scope. It is in this milieu of interdisciplinary renaissance and revolution that individual disciplines fulfill their true creeds, identities, and purposes. The complete overcoming of STEM challenges is wholly predicated on the complete understanding of the disciplinary sub-components of these challenges and then on how these sub-components relate to, differ from, and influence each other. This is the essence of interdisciplinary learning, knowledge, and authentication.

Twenty-first century challenges will require bold, innovative solutions that are knitted and seamlessly interwoven in the fabric of connective, synthesized learning that draws upon knowledge from across the disciplines. These novel solutions will come from critical thinkers who are flexible and reflective students of interdisciplinary learning and who grasp and are cognizant of the interplay between—and the dependency within—disciplines as problem-solving is being conducted. Philosopher Karl Popper understood this well. He emphasized the assertion that since problems are unbounded by disciplines, so too must their solutions be.[1] However, for these solutions to be interdisciplinary authentic, Biox and Duraising argue that the student problem-solvers: (1) *be well grounded in the disciplines*: show rigorous understanding and use of disciplinary tools, perspectives, and approaches, (2) *show critical awareness*: be mindful of the purpose and the means by which the disciplines have been brought together—the discipline's potential contributions and limitations, and (3) *exhibit advanced understanding*: demonstrate that they have developed a new model, perspective, insight, or solution that could only be possible by integrating more than one discipline.[2] As these students engage in authentic interdisciplinary learning, they not only increase cognitive abilities[3] and problem-solving skills, but they also acquire new perspectives, engage in deeper more comprehensive probing and understanding of issues and problems, devise alternate solutions, and start on their way to becoming lifelong learners. Moreover, interdisciplinary learning also allows them to see global and local challenges through a different set of lenses. These interdisciplinary lenses enable them to analyze, synthesize, and assimilate pre-existing ideas with new critical thinking abilities to develop new, transformative knowledge and opportunities. To this end, as described below, geoscience research (an ideal STEM platform for interdisciplinary learning) was used as a catalyst for interdisciplinary learning among undergraduate students at the City University of New York.

PROMOTING THE INTERDISCIPLINARY NATURE OF THE GEOSCIENCES

A plethora of recent studies gives both evidence of and solutions to the nation's STEM crisis.[4] Perhaps unlike any other STEM discipline, geoscience and its related subfields (e.g., environmental earth science, atmospheric science, oceanography, hydrology, geology, geochemistry, and geophysics, among others), are uniquely and inherently interdisciplinary. However, they are also fraught with many of the problems—from climate change impacts and natural disasters to renewable energy sources—that still need twenty-first century solutions. The geosciences, therefore, offer both challenges and opportunities that range from the neighborhood scale to the global scale. Interdisciplinary ingenuity is needed to tackle these challenges and to exploit, advance, and realize the opportunities they provide.

For these endeavors, students must be trained and equipped with interdisciplinary tools and skills. A National Science Foundation (NSF) Research Experiences for Undergraduates (REU) grant was used in part to demonstrate that students who are polarized and isolated in their individual STEM disciplines can be trained and equipped to think, act, and produce in an interdisciplinary manner via structured engagement in a high-impact, best practices-driven undergraduate geoscience research program. This program affords students the opportunity to participate in state-of-the-art satellite- and ground-based remote sensing of the cryosphere, the lithosphere, the hydrosphere, the biosphere, and the atmosphere.

The Interdisciplinary Geoscience Undergraduate Research Experience

New York City College of Technology (City Tech) was awarded two NSF REU grants from 2008 to 2015. The REU program was intentionally designed to encourage and to foster interdisciplinary learning, particularly in the geosciences through satellite- and ground-based remote-sensing projects. REU scholars were actively engaged in full-time research for nine weeks in the summer and one day per week in the fall and spring semesters.

The learning outcomes for the REU scholars were established by using the benefits gained statements from the evaluation instrument: Undergraduate Research Student Self-Assessment Survey.[5] These indicated gains are potentially portable within and beyond their STEM disciplines. REU scholars in the one-year research program were expected to report gains in the following four areas:

1. *Thinking and working like a scientist*
   REU scholars will have a clear understanding of how the scientific disciplines are connected and of how scientific research is done. They will be able to comprehend and apply problem-solving and analytical skills to their research. Furthermore, their knowledge of the geosciences will be expanded and strengthened.

2. *Personal and professional gains related to research work*
   REU scholars will display confidence conducting research, working collaboratively with other researchers, and excelling in future science courses.

3. *Becoming a scientist*
   REU scholars will have the skills and ability to work independently and to reflect and display ownership of their own research.

4. *Skills*
   REU scholars will have the ability to write scientific papers, explain their research to others in the field and to broader audiences through research presentations, prepare scientific poster presentations, articulate the relevance of their research, and know and understand the existing body of research relevant to their topic.

The REU program design supported the interdisciplinary learning of geoscience concepts in four areas of the research process: (1) mentoring, (2) mini-courses and seminars, (3) field experiences, and (4) conference participation.

(1) *Mentoring*

The REU program implemented a structured mentoring paradigm that consisted of an interdisciplinary team. Each REU scholar was affiliated with a team that involved a faculty research scientist, a post-doctoral research scientist, a graduate student, and a high-school student. The faculty research scientist designed the research project with input from the post-doctoral research scientist and the graduate student. Although they were all working in the same laboratory, each person differed in expertise. Therefore, an interdisciplinary team culture was created. Resources, expertise, and experiences were shared and a community of practice was formed. As a result, each REU scholar benefited from the relationships developed. Additionally, the REU scholar also became a mentor to a high-school student who also benefitted from the interdisciplinary team structure.

(2) *Mini-Courses and Seminars*

REU scholars come to the REU program with varied educational skill sets, and they often do not have the necessary background for research, especially in the geosciences. To prepare them with the tools needed for satellite- and ground-based remote-sensing research, the REU program organized four crucial mini-courses: MATLAB, remote sensing, geographic information systems (GIS), and basic statistics. These mini-courses were offered during the second and third weeks of the nine-week summer program.

MATLAB: This mini-course introduced students to the fundamentals of MATLAB programming. Students were taught how to use MATLAB as a tool to study and analyze remote-sensing data and to manipulate basic algorithms for remote-sensing applications. The software's statistical, graphics, mapping, and visualization tools were applied to real-time data sets from both satellite- and ground-based measurements.

Remote sensing: This mini-course provided students with the basic knowledge necessary to begin to understanding the key fundamentals of the science of environmental remote sensing and its related geoscience applications. Students learned about environmental measurements obtained from space platforms and from the variety of other platforms that are used for remote-sensing applications. They were introduced to polar orbiting satellites, geostationary satellites, active and passive systems, the atmosphere and atmospheric sounding techniques, interferometric and LIDAR systems, image processing, and radiative transfer.

Geographic information systems (GIS): Essential GIS concepts and techniques, including GIS database queries, tabular data manipulation, spatial and attribute data editing, data presentation via maps and charts, map layers, area measurement, scale, and symbology were taught. The relevance and connectivity of GIS to remote sensing were highlighted.

Basic statistics: In this mini-course, basic descriptive statistics were taught with special emphasis on correlation and linear regression techniques using real world applications. Taken together as an interdisciplinary package, these mini-courses provided the REU scholars the skills and the basic knowledge needed for their geoscience research projects.

Seminars: Another important component of the REU program is its seminars. Faculty members participating in the REU program were renowned scientists from the Remote Sensing Science and Technology Center (CREST) located at the City College of New York. CREST is a National Oceanic and Atmospheric Administration (NOAA) research facility. NOAA-CREST conducted a weekly seminar series (at the City University of New York and all partner institutions), by inviting geoscience experts—the majority of whom were from NOAA line offices—to present research projects that were congruent to NOAA-CREST sciences. These seminars were largely attended by NOAA-CREST scientists, their graduate and undergraduate students, and the research community within each campus. The seminar series provided students with a broader perspective of their research projects, and the seminars helped to initiate and to strengthen research collaborations. Moreover, they also increased scientific horizons beyond the students' individual research areas.

(3)   *Field Experiences*

In order to establish geoscience research career pathways, to expand student exposure to remote sensing beyond the campus, and to promote the attainment of graduate degrees, field experiences were organized at different research and educational sites. The opportunities were arranged so that the REU scholars would have an opportunity to tour the facilities and to meet scientists with interdisciplinary academic backgrounds at the laboratory.

These field experiences inspired the students to keep aspiring toward STEM success, and they provided the students with a vision of what they can become after they graduate from college. The students were exposed to and interacted with scientists who had degrees and interdisciplinary experiences that supported and advantaged them in their current positions.

Without establishing field and site partnerships, exposing the REU scholars to careers beyond their own disciplines would not have been possible. The sites below provided the following benefits:

- *The Brookhaven National Laboratory*—Weather balloon launches at the National Weather Service were observed. These radiosonde launches were interdisciplinary in nature, as they involved teams of scientists with varying STEM degrees who used the collected data to study various aspects of the troposphere.
- *The National Center for Weather and Climate Prediction*—The center provided REU scholars with lectures and tours. Opportunities to interact with the operational scientists who were on duty monitoring and predicting atmospheric and oceanic dynamics provided an engaging experience. These experiences were interdisciplinary as meteorologists, oceanographers, and computer scientists converged to study and share their integrative, interdisciplinary knowledge about the Earth system in real-time, operational, and collaborative sessions.
- *The American Museum of Natural History*—Climate change and astronomy exhibits were highlighted at the museum. Informal, interdisciplinary learning occurred as resident scientists helped students to make the system-wide connections and feedback mechanisms between the different inter-connected strands of the geosciences.
- *The Cary Institute of Ecosystem Studies*—Forums on translational ecology and hydrology were provided. Interdisciplinary hydro-ecological lessons were taught by the geoscientists and the researchers at the Institute.
- *The Solar Panel House*—Demonstrations of how solar energy can be used to provide 100 % of the energy needs of a house were presented. Mechanical and civil engineering faculty demonstrated how seamlessly the interdisciplinary relationships between engineering and the geosciences are with regards to the Earth's ultimate energy source (the sun), renewable energy, economics, and living conditions.

(4) *Conference Presentations*

The culminating interdisciplinary geoscience component of the REU program is the presentation of the students' research projects at national and regional geoscience and STEM conferences such as the American Geophysical Union (AGU), the American Meteorological Society (AMS), the NOAA Educational Partnership Program, the Louis Stokes Alliance for Minority Participation, the National Organization for the Professional Advancement of Black Chemists and Chemical Engineers, the National Science Foundation (Emerging Researchers National Conference in STEM), and the Mathematical Association of America (MAA). This additional level of exposure and scholarly validation is extremely valuable and beneficial to the undergraduates, because it allows them to participate in the geoscience realm in ways that authenticate their interdisciplinary experience. They feel a sense of purpose and belonging as they join, exchange, and network with fellow students and scientists. Vistas of graduate school and career options become open, and the REU scholars are inspired to become the future geoscientists that they could not envision being before joining this interdisciplinary program.

REU STUDENT AND FACULTY PARTICIPANTS

The REU geoscience research program encourages students majoring in any STEM discipline to apply. With a shortage of students majoring in the geosciences, the premise of the research opportunity is to create awareness of how interdisciplinary the geosciences really are and to provide a pathway for students to acquire sufficient geoscience knowledge that they may afterward apply to the geoscience workforce with their STEM majors. The REU scholars were recruited and selected from several of the twenty-three campuses of the City University of New York. Students were primarily in their junior or senior year. However, a few exceptional sophomores from community colleges and two-year programs have participated in the program because of their prior research background. The students must also be either US citizens or permanent residents with a grade point average of 3.0/4.0 or above. Being a geoscience major or having a geoscience background is not a requirement for the REU program.

Since the awarding of the grant in 2008, a total of 78 students have successfully completed the year-long REU program. Of the 78 students, 82.1 % (64) majored in a STEM discipline other than the geosciences, and only 17.9 % (14) majored in the geosciences. The REU scholars spanned 29 different STEM majors (Table 6.1). Among the 78 students, 42.3 % (33) of them identified themselves as African-American (non-Hispanic), 20.5 % (16) as Hispanic, 20.5 % (16) as Asian/Pacific Islander, 15.4 %

Table 6.1 REU scholars by STEM majors

Major (N = 78)	Number of students (%)
Architectural Technology	1 (1.3 %)
Biology	1 (1.3 %)
Biology Engineering	1 (1.3 %)
Biomedical Informatics	1 (1.3 %)
Civil Engineering	7 (9.0 %)
Computer Engineering	1 (1.3 %)
Computer Engineering Technology	2 (2.6 %)
Computer Information Systems	1 (1.3 %)
Computer Science	6 (7.7%)
Computer Systems Technology	2 (2.6 %)
Earth and Atmospheric Sciences	5 (6.4 %)
Electrical Engineering	8 (10.3 %)
Engineering Science	2 (2.6 %)
Environmental Control Technology	1 (1.3 %)
Environmental Engineering	8 (10.3 %)
Forensic Science	1 (1.3 %)
General Science	1 (1.3 %)
Information Systems Management	1 (1.3 %)
Liberal Arts and Sciences	2 (2.6 %)
Mathematical Science	5 (6.4 %)
Mathematics (Applied)	5 (6.4 %)
Mathematics Education	1 (1.3 %)
Mechanical Engineering	6 (7.7 %)
Mechanical Engineering Technology	2 (2.6 %)
Medical Laboratory Science	1 (1.3 %)
Meteorology	1 (1.3 %)
Neuroscience	1 (1.3 %)
Physics	3 (3.8 %)
Telecommunications Engineering Technology	1 (1.3 %)

(12) as Caucasian, and 1.3 % (1) as other. The program had 64.1 % (50) males and 35.9 % (28) females. Of the total cohort, 83.3 % (63) of the participants were considered underrepresented minorities in the STEM disciplines (defined as Blacks or African-Americans, Hispanics or Latinos, American Indians or Alaskan Natives, Native Hawaiians or other Pacific Islanders, or females).

The students participated in a range of interdisciplinary research projects in satellite- and ground-based remote sensing. These interdisciplinary research projects focused on the study of the earth's atmosphere, hydrosphere, cryosphere, biosphere, and lithosphere. The projects included the study of hurricanes, atmospheric water vapor distribution, soil moisture, vegetation, sea ice, air quality, coastal waters, and climate change. REU scholars from among the disciplines in Table 6.1 successfully completed and presented their geoscience research projects at numerous national, regional, and local conferences across the country. Many of them have won awards at these and other geoscience and STEM conferences. Sample research project titles by student majors are shown in Table 6.2.

Table 6.2 Sample research project titles by student majors

Student majors	Research project titles
Architectural Technology	• Exploring DMSP (Defense Meteorological Satellite Program) SSM/T2: Measurements to understand atmospheric water vapor distribution
Biomedical Informatics	• Spatial variability of ambient ozone concentrations during three heat waves in the northeast megaregion of the USA
Civil Engineering	• Developing an algorithm to investigate cloud lifetime using GOES satellite thermal channel information
Computer Engineering	• Combining Spatial Kriging with satellite estimates to obtain a regional estimation of PM2.5
Computer Engineering Technology	• Observing aerosol mass densities during the trans-Atlantic transport of Saharan dust and biomass burning aerosols
Computer Science	• Flood prediction using multidimensional analysis of precipitation and inundation in the Mekong river delta
Earth and Atmospheric Science	• Creating water body maps for the Pacaya Samiria, the everglades, and the US gulf coast using NASA UAVSAR imaging radar data
Electrical Engineering	• Use of reconnaissance aircraft data in estimating hurricane intensity
Environmental Engineering	• Modeling optical properties of aerosols using microphysical retrievals from air quality models
Information Systems Management	• Band 6 restoration for snow mask: Yellowstone case study
Liberal Arts and Sciences	• Detection of land cover change and drought trend using brightness temperature and microwave emission
Mathematics (Applied)	• Remote sensing of ice in the Caspian sea via MATLAB programming
Mechanical Engineering	• Preliminary analysis: electricity consumption changes in California
Mechanical Engineering Technology	• Use of satellite images for surface conditions monitoring in the upper Mississippi watershed during the flood event of 2008
Physics	• Connecting air pollution parameters to optically measured parameters to assess air quality monitoring capabilities

Not only were the REU scholars' projects interdisciplinary in nature, but their faculty mentors themselves were also from a variety of disciplines and interdisciplinary research fields. Faculty members participating in the REU program were renowned scientists from CREST located at the City College of New York. CREST is a NOAA research facility that specializes in satellite- and ground-based remote-sensing research. Faculty mentors were also from CREST affiliates formed with City Tech's Remote Sensing and Earth System Sciences (ReSESS) Center and the Bronx Community College's Geospatial Center. Faculty mentors were also from NOAA line offices. Research foci ranged from developing satellite multi-sensor rainfall and snowfall retrieval algorithms and merging model estimates with ground truth measurements to improve quantitative precipitation estimation (QPE) of the construction of global aerosol, cloud, and trace-gas climatologies derived from satellite measurements. A list of some research expertise of faculty mentors by department is found in Table 6.3.

The results from completing the one-year REU program have been generally positive. Of the 78 REU scholars who have participated in the program, 18.8 % (15) of them are in graduate school in the STEM disciplines, 16.3 % (13) of them have graduated and are now employed in the STEM workforce, and 65 % (51) of them are continuing with their STEM majors. Moreover, four of the REU scholars have won first-place recognition for their research at national and regional conferences, another four have won second-place recognition at regional and local conferences, and three other REU scholars are co-authors for two peer-reviewed publications and one book chapter. The REU program claims a 100% STEM retention rate for its REU scholars.

The REU scholars evaluated their experience by completing selected questions from the Undergraduate Research Student Self-Assessment survey.[6] The focus of the survey was that a standard set of potential learning gains from the one-year research experience was collected for evaluation from each student respondent. Of the 78 students, 37 responded to the survey. The results in Table 6.4 showed positive student-reported gains in the following four areas:

1. *Thinking and working like a scientist*
   REU scholars reported "good" to "great" gains in understanding how connected or interdisciplinary the various scientific disciplines really are to each other. They gained understanding of how science research is conducted; thus, their analytical skills in identifying pat-

Table 6.3 Faculty research expertise by STEM departments

STEM department	Research expertise
Civil Engineering	• Water resources and land hydrology: development of advanced techniques for monitoring of hydrometeorology from passive and active microwave satellite observations
Construction Management and Civil Engineering	• Civil engineering and water resources: using satellite information in climate studies, vegetation structure profile, soil moisture estimation, snow cover prediction, precipitation, and estimation of soil freeze/thaw state
Earth and Atmospheric Sciences	• Meteorology: satellite remote sensing and clouds in the context of climate and climate change
Electrical Engineering	• Improvement of methods for using current and future satellite measurements in air quality forecasting (MODIS, MISR, CALIPSO, APS, VIIRS, GOES-R, etc.)
	• Tropospheric remote sensing and air quality (TRAQ): satellite algorithm development and validation, ground-based remote-sensing network, ground-based in-situ measurements, sampling and sample analysis and speciation, modeling and validation, and health impacts
	• Remote sensing of coastal waters: evolution of measurements approaches for coastal water parameters, development of suite of field measurement capabilities for algorithm testing and satellite validation in coastal waters, and improvement/development of algorithm for remote sensing of coastal waters
Electrical Engineering Technology	• Optics and remote sensing of the atmosphere as well as model development and model validation using land- and space-based remote-sensing systems. Specializes in modern optical metrology systems used in noncontact measurements of space instruments
Mechanical Engineering	• Energy, sustainability, climate change, climate modeling, and remote sensing: recognized in the field of urban climatology, remote sensing, and renewable energy
Physics	• Tropical meteorology: specializes in the use of satellites to study hurricanes, climate and climate change impacts, urban climate and micro-meteorology, and remote-sensing applications to air pollution and hydrometeorology

terns and interpretation skills of the results generated were increased. Moreover, these students indicated higher gains in expanding and grounding their knowledge of the geosciences from the REU experience.

2. *Personal and professional gains related to research work*

"Good" to "great" gains were reported in the REU scholars' confidence in their ability to do well in future science courses and

Table 6.4 Means and standard deviations of responses regarding benefits gained from the REU research experience

Benefits gained from the REU research experience: 1 = no gain, 2 = a little gain, 3 = good gain, 4 = great gain	Mean (SD); N = 37
Gains in thinking and working like a scientist	
Understanding how science research is done	3.58 (0.55)
Understanding how to collect scientific data	3.38 (0.79)
Problem-solving in general	3.35 (0.79)
Analyzing data for patterns	3.47 (0.61)
Interpreting results from analyzing scientific data	3.31 (0.75)
Identifying limitations in research methods and designs	3.16 (0.93)
Understanding the connections among scientific disciplines	3.39 (0.65)
Extending my knowledge	3.69 (0.58)
Solidifying my knowledge	3.50 (0.74)
Personal gains related to research work	
Confidence in my ability to do research	3.41 (0.83)
Confidence in my ability to contribute to science	3.36 (0.80)
Comfort in working collaboratively with others	3.42 (0.73)
Confidence in my ability to do well in future science courses	3.58 (0.65)
Gains in becoming a scientist	
Ability to work independently	3.36 (0.64)
Understanding what everyday research work is like	3.39 (0.73)
Gains in skills	
Writing scientific reports or papers	3.08 (0.87)
Making oral presentations	3.44 (0.69)
Defending an argument when asked questions	3.24 (0.78)
Preparing a poster	3.58 (0.55)
Working with computer software	3.53 (0.61)
Understanding journal articles	2.92 (0.80)

      contribute to science, to conduct research, and to work in a collaborative setting.

3. *Becoming a scientist*

    REU scholars showed "good" to "great" gains in their ability to work independently and in their understanding of what doing and conducting research is like as a scientist.

4. *Skills*

    Higher gains were reported in the REU scholars' ability to present their research orally, prepare a poster presentation, and use MATLAB in analyzing their data. They also indicated "good" gains in their ability to write scientific papers and understand scientific journal articles.

In addition, questions regarding benefits gained from the various REU program components were solicited from the students (Table 6.5). With respect to the mini-courses, the REU scholars reported higher learning gains for MATLAB followed by statistics. Reasonably, the MATLAB and statistical abilities gained by the REU scholars are particularly portable and valuable for those majoring in a STEM discipline. For the mentoring component, the survey showed that REU scholars valued their graduate or post-doctoral research scientist mentors over their faculty scientist mentors. Because most of the research time was spent under the direction and guidance of their graduate or post-doctoral research scientist mentors, the REU scholars indicated greater learning gains from them. However, the REU scholars found "little" to "good" gain in mentoring their assigned high-school students. Overall, there were "good" gains from the community of REU scholars.

The REU scholars found the geoscience exposure trips rewarding. Through their interactions with the scientists, students discovered how interdisciplinary the scientists were in their expertise and in the positions they occupied. This awareness was particularly noted when the students

Table 6.5 Means and standard deviations of responses regarding the REU program components

Benefits gained from REU program components: 1 = no gain, 2 = a little gain, 3 = good gain, 4 = great gain	Mean (SD); N = 37
Mini-courses	
MATLAB mini-course	3.24 (0.99)
Remote-sensing mini-course	2.91 (0.89)
GIS mini-course	2.88 (0.91)
Basic statistics mini-course	2.97 (0.91)
Remote-sensing orientation seminars	3.00 (0.91)
Mentoring	
Faculty scientist mentoring	3.08 (0.87)
Graduate student/post-doctoral research scientist mentoring	3.26 (0.99)
High-school mentee	2.70 (0.93)
Community of REU scholars	3.14 (0.93)
Field experiences	
American Museum of Natural History	2.97 (0.87)
Brookhaven National Laboratory	3.35 (0.71)
National Weather Service	3.41 (0.71)
Cary Institute	3.15 (0.99)
NOAA Center for Weather and Climate Prediction	3.67 (0.59)

visited the NOAA Center for Weather and Climate Prediction and the National Weather Service. The REU scholars also highly valued their participation at national, regional, and local conferences and events.

To capture students' perspectives (expectations, experience, knowledge, ability, exposure to research, future goals, etc.) of the REU program, a series of formative and summative assessment surveys were also conducted. The survey results revealed the following:

- The mini-courses in MATLAB, GIS, and remote sensing were extremely effective in providing the STEM students with the necessary foundation for interdisciplinary satellite- and ground-based remote-sensing research.
- The mini-course in statistics was useful and applicable not only for their geoscience research, but also for their own STEM majors.
- Due to their participation in the REU program, most of the students plan to pursue STEM graduate degrees.
- On average, REU scholars each made about six presentations of their interdisciplinary REU geoscience research at national, regional, or local conferences; by so doing, the survey results indicate that their oral and poster presentation skills have tremendously improved.
- Due to their participation in the REU program, REU scholars have gained a greater motivation for interdisciplinary learning.
- Their participation in the REU program has made them independent, interdisciplinary thinkers while simultaneously enabling them to function well within a community of learners where teamwork is essential.
- Due to their participation in the REU program, the likelihood of the students pursuing both Master's and PhD degrees in the geosciences (satellite- and ground-based remote sensing) significantly increased.
- REU scholars now feel that they understand better how to think critically like geoscientists do.
- After the one-year REU experience, REU scholars felt prepared for more demanding geoscience research endeavors, and their analytical and data management skills have improved.
- Many of the REU scholars enjoyed their interdisciplinary, geoscience mentoring experience—both by being mentored by graduate students and by providing guidance to their high-school mentees.

Reflections on the REU Experience

A number of REU scholars reflected on their experiences within the course. For example, one Mathematics major stated, "I gained valuable programming skills which made my mathematical modeling classes easier. I gained public speaking skills which were essential in class presentations. Lastly, I was able to experience a working application of the concepts I studied." Likewise, a Biomedical Informatics major remarked that the REU program "gave me more computational skills and research tools, and [it has] allowed me to use those skills for my biomedical informatics program." In addition, a Mechanical Engineering major credited the course for its improving reasoning skills: "The REU program taught me to think critically and beyond information provided to you. This is because research requires one to think beyond what has already been done. The professor I worked with always pushed me to think this way and to maintain a positive outlook despite the challenges. This made my academic class work seem like a breeze!" Another student remarked on the way the course helped students think in interdisciplinary ways that will help them in their future careers: "As a Computer Science major, I was primarily in the software development aspect of the REU. I was able to hone my skills for the industry while working with other researchers and scientists to develop tools to make research easier. This broadened my understanding of other areas in the industry knowing that my major could be involved with different majors and backgrounds and creating collaboration." Another Biomedical Informatics major also hailed the way the course promoted interdisciplinary thinking: "I really appreciated this opportunity to embark on a journey to broaden my understanding of the sciences in general by working in close cooperation with a professional in the field on a discipline in which I had no previous knowledge. I feel that I have grown both as an individual and as a professional."

In addition to these reflections, other programmatic assessments reveal that many of the underrepresented minority students in the program would not have continued down the pathway toward advanced degrees were they not trained and provided with the interdisciplinary geoscience experiences of the program. The interdisciplinary training they received boosted their confidence, increased their critical thinking skills, and improved their understanding of the environmental complexities for which they seek solutions. Members of the full cohort of REU scholars have now joined the ranks of a new cadre and corps of interdisciplinary learners and practitioners who are equipped and trained with the necessary skills sets, confidence, and ability to tackle the many interdisciplinary challenges that exist and that will emerge throughout this century. They have been taught how to approach, investigate, and interrogate with the interdisciplinary tools and mindset that the REU program gave to and developed in them.

CONCLUSION

As the need for interdisciplinary learning becomes more of a national imperative, this REU program demonstrates that a comprehensive geoscience research experience given to STEM majors at the appropriate academic juncture can be a useful and efficient method to bring about critical interdisciplinary outcomes. Since twenty-first century problems have become less discipline-specific and more interwoven across disciplinary boundaries, careful and intentional knitting of different disciplinary strands can produce the integrative tools necessary to confront complex problems. Knitting various strands of STEM disciplines, for example, can produce interdisciplinary ingenuity and innovation that not only excites STEM students, but also equips them with the skills and the confidence needed to analyze, understand, and seek solutions to problems of many scales and of many resolutions. The tasks and the problems of the twenty-first century are challenging and daunting, and any amelioration of them must be sought via the mechanisms and methodologies that are inherent and sui generis to interdisciplinary scholarship; this type of scholarship needs to be embraced by institutions of learning and made available to students now, not later. National and global security and advancement hinge on this pivotal scholastic paradigm.

Finding solutions to the difficult, multi-pronged problems that face humankind at all levels and in all sectors will require bold new pedagogic initiatives that are needful, practical, and make simple common sense. Popper is correct: if problems have no disciplinary borders, then their solutions must mimic their disciplinary un-restrictedness. To this end, the following six recommendations are offered. First, educators should acknowledge that complex problems (and their solutions) are often not hemmed in by disciplinary boundaries. Second, understand the individual components (disciplines) involved in the problem before attempting to delineate the relationships (interdisciplinary nature) between them. Third, promote interdisciplinary learning and its many advantages. Fourth, introduce, expose, and engage students in interdisciplinary learning early and often. Fifth, set clear, achievable, and measurable interdisciplinary student learning goals. Finally, expand the experiment of this study to non-STEM disciplines.

Consequently, for formidable societal challenges to be met, and for learning to be truly optimized, a philosophical paradigm shift toward inter-disciplinary pedagogy is needed among the educational cauldrons in which knowledge brews and from which knowledge is dispensed. Institutions of learning will need to embrace this paradigm shift if they are to remain relevant and contributory to global advancement. Many institutions will have to revive, reconfigure, and reimagine extant programs and courses and simultaneously introduce new ones so as to attract and educate a new and growing company of revolutionary, interdisciplinary thinkers and scholars. Old frameworks and dull, unimaginative, pedantic schemes of packaging and delivering the product of knowledge will soon be abandoned and discarded, for they are inadequate and ill-designed to meet the needs of a rapidly advancing and interdependent local, regional, national, and global populace. Interdisciplinary courses, interdisciplinary research projects, and interdisciplinary professional development training for faculty will feature prominently in this new and necessary paradigm. Fresh, interdisciplinary perspectives promise to invigorate and enliven both faculty and students and thereby transform academic institutions into continuous, sustainable beehives of interdisciplinary activities that are replete with practical challenges that will both require and produce a hub of innovation, discovery, and intellectual stimuli. The time has now come for a widespread, far-reaching, interdisciplinary renaissance to open and to expand a critical mode of acquiring and dispensing knowledge. Therefore, let the potential of mutual symbioses of disciplines be explored; let interdisciplinary learning be a formidable tool to subdue and overcome twenty-first century challenges, and let the new renaissance begin!

NOTES

1. Karl. R. Popper, *Conjectures and Refutations: The Growth of Scientific Knowledge* (New York: Routledge and Kegan Paul, 1963).
2. Mansilla Biox and Elizabeth Dawes Duraising, "Targeted Assessment of Students' Interdisciplinary Work: An Empirically Grounded Framework Proposed," *Journal of Higher Education* 78, no. 2 (2007): 215–237.

3. Repko Allen, "Assessing Interdisciplinary Learning Outcomes," Working Paper, School of Urban and Public Affairs, University of Texas at Arlington. 2009.

4. Domestic Policy Council, American Competitiveness Initiative: Leading the World in Innovation, 2006, accessed May 10, 2015, http://georgewbush-whitehouse.archives.gov/stateoftheunion/ 2006/aci/aci06-booklet.pdf; Institute of Medicine, National Academy of Science, and National Academy of Engineering, *Rising Above the Gathering Storm: Energizing and Employing America for a Brighter Economic Future* (Washington, DC: National Academies Press, 2007); Educate to Innovate Campaign for Excellence in Science, Technology, Engineering and Math (STEM) Education [Press release], November 23, 2009, accessed May 2015, http://www.whitehouse.gov/the-press-office/president-obama-launches-educate-innovate-campaign-excellence-science-technology-en; President's Council of Advisors on Science and Technology, *Report to the President—Engage to Excel: Producing One Million Additional College Graduates with Degrees in Science, Technology, Engineering, and Mathematics*, February 2012, accessed May 15, 2015, http://www.whitehouse.gov/sites/default/files/microsites/ostp/pcast-engage-to-excel-final_2-25-12.pdf; American Geological Institute, "Chapter 2: Four-year Colleges and Universities," *Status of the Geoscience Workforce*, 2009, accessed June 15, 2015, http://www.americangeosciences.org/sites/default/files/2009-FourYrInstitutions_rev082509.pdf; S. James Gates, Jr. and Carl Mirkin, "Encouraging STEM Students is in the National Interest," *Chronicle of Higher Education*, June 25, 2012, accessed May 2, 2015, http://chronicle.com/article/Encouraging-STEM-Students-Is/ 132425; Reginald Blake, Janet Liou-Mark, and Chinedu Chukuigwe, "An Effective Model for Enhancing Underrepresented Minority Participation and Success in Geoscience Undergraduate Research," *Journal of Geoscience Education* 61, no. 4 (2013): 405–414; Reginald A. Blake, Janet Liou-Mark, and Reneta D. Lansiquot, "Promoting the Geosciences among Grades 8–12 Minority Students in the Urban Coastal Environment of New York City," *Journal of Geoscience Education* 63, no. 1 (2015): 29–40.

5. Undergraduate Research Student Self-Assessment, "Ethnography and Evaluation Research," University of Colorado at Boulder, Boulder, CO, 2009, accessed June 27, 2015, http://www.salgsite.org

6. Ibid.

Acknowledgements The REU program was supported by the National Science Foundation Research Experiences for Undergraduates Grants No. 0755686 and No. 1062934, under the direction of Dr. Reginald A. Blake, Principal Investigator, and Dr. Janet Liou-Mark, Co-Principal Investigator. The authors are solely responsible for the content of this article, and it does not necessarily represent the views of the NSF or of NOAA-CREST. The authors would like to thank Dr. Reza Khanbilvardi and Dr. Shakila Merchant at the NOAA-CREST Center at the City College of New York sincerely for their dedication to the REU program from its very inception. The authors also wish to thank the committed CREST faculty research scientists, post-doctoral research scientists, and graduate students for their leadership in this critical and rewarding program. The authors are grateful to Mr. Chinedu Chukuigwe and Ms. Laura Yuen-Lau for their assistance in sustaining a viable undergraduate research program that enhanced students' interdisciplinary learning skills, and to Mr. Ricky Santana for assisting in the data analyses. To all the REU scholars, the authors would like to express their gratitude for your willingness to develop your interdisciplinary skills by acquiring the foundational knowledge necessary for the geoscience research, persisting in the understanding of the scientific procedure, and tolerating and overcoming the obstacles faced in the research process. The program and its success would certainly be impossible without the unwavering dedication of these champions.

BIBLIOGRAPHY

Allen, Repko. 2009. "Assessing Interdisciplinary Learning Outcomes." Working Paper, School of Urban and Public Affairs, University of Texas at Arlington.

American Geological Institute. 2009. "Four-year Colleges and Universities." Chap. 2 in *Status of the Geoscience Workforce*. Accessed June 15, 2015. http://www.americangeosciences.org/sites/default/files/2009-FourYrInstitutions_rev082509.pdf

Biox-Mansilla, Veronica, and Elizabeth Dawes Duraising. 2007. "Targeted Assessment of Students' Interdisciplinary Work: An Empirically Grounded Framework Proposed." *Journal of Higher Education* 78 (2): 215–237.

Blake, Reginald A., Janet Liou-Mark, and Chinedu Chukuigwe. 2013. "An Effective Model for Enhancing Underrepresented Minority Participation and Success in Geoscience Undergraduate Research." *Journal of Geoscience Education* 61 (4): 405–414.

Blake, Reginald A., Janet Liou-Mark, and Reneta D. Lansiquot. 2015. "Promoting the Geosciences among Grades 8–12 Minority Students in the Urban Coastal Environment of New York City." *Journal of Geoscience Education* 63 (1): 29–40.

Domestic Policy Council. 2006. *American Competitiveness Initiative: Leading the World in Innovation*. Accessed May 10, 2015. http://georgewbush-whitehouse.archives.gov/stateoftheunion/2006/aci/aci06-booklet.pdf

Educate to Innovate Campaign for Excellence in Science, Technology, Engineering and Math (STEM) Education [Press release]. November 23, 2009. Accessed

May 2, 2015. http://www.whitehouse.gov/the-press-office/president-obama-launches-educate-innovate-campaign-excellence-science-technology-en

Gates, James S., Jr., and Carl Mirkin. "Encouraging STEM Students is in the National Interest." *Chronicle of Higher Education.* June 25, 2012. Accessed May 2, 2015. http://chronicle.com/article/Encouraging-STEM-Students-Is/132425

Institute of Medicine, National Academy of Science, and National Academy of Engineering. 2007. *Rising Above the Gathering Storm: Energizing and Employing America for a Brighter Economic Future.* Washington, DC: National Academies Press.

Popper, Karl R. 1963. *Conjectures and Refutations: The Growth of Scientific Knowledge.* New York: Routledge and Kegan Paul.

President's Council of Advisors on Science and Technology. "Report to the President—Engage to Excel: Producing One Million Additional College Graduates with Degrees in Science, Technology, Engineering, and Mathematics." February 2012. Accessed May 15, 2015. http://www.whitehouse.gov/sites/default/files/microsites/ostp/pcast-engage-to-excel-final_2-25-12.pdf

Undergraduate Research Student Self-Assessment. 2009. "Ethnography and Evaluation Research." *University of Colorado at Boulder, Boulder, CO.* Accessed June 27, 2015. http://www.salgsite.org

CHAPTER 7

Conclusion: Integrating Interdisciplinary Pedagogies

Reneta D. Lansiquot and Tamrah D. Cunningham

Abstract Synthesizing the best practices and lessons learned from collaborative interdisciplinary classrooms, this chapter offers a retrospective of student experience, discussing how this pedagogical strategy effectively promotes self-authorship, manifested through the learners' ability to reflect and base judgments on their knowledge and interdisciplinary understanding, as well as their ability to integrate multiple disciplines to accomplish a task. Free educational technologies were used to scaffold student learning via, for example, place-based learning in virtual worlds, using such features as integrating digital concept maps and three-dimensional virtual worlds. Despite the variability and unpredictability of individual experiences, student perspectives provide evidence for the unique challenges and distinct advantages of team-taught interdisciplinary courses.

Keywords Concept maps • Co-teaching • Educational technology • Interdisciplinary pedagogies • Interdisciplinary understanding • Virtual worlds

R.D. Lansiquot (rlansiquot@citytech.cuny.edu ✉)
Department of English, New York City College of Technology,
City University of New York, Brooklyn, NY, USA

T.D. Cunningham
Department of Game Design, Tisch School of the Arts,
New York University, Brooklyn, NY, USA

© The Editor(s) (if applicable) and The Author(s) 2016
R.D. Lansiquot (ed.), *Interdisciplinary Pedagogy for STEM*,
DOI 10.1057/978-1-137-56745-1_7

127

Interdisciplinary competence should be an integral part of undergraduate education. Team-taught interdisciplinary courses, wherein the different perspectives are provided, help students make vital connections between courses. The preceding chapters of this book included best practices and lessons learned from collaborative interdisciplinary classrooms, both formal and informal. The best practices include student collaboration, encouraging students to consider of multiple perspectives, and having students analyze critically the literature on a topic, online discussions, and survey data. Such tasks focus on the process of integrating insights from disparate disciplines using creativity, critical thinking, and technology-supported collaborative learning.

As one guest lecturer in the team-taught interdisciplinary course, Weird Science: Interpreting and Redefining Humanity, has observed, integration is a necessary component of interdisciplinary studies, especially the process in which the emerging learner becomes skilled at integrating diverse disciplines; this process has been elucidated by Gadamer, who described a *sensus communis*, wherein all disciplinary knowledge is the outgrowth of a more fundamental acculturation process that begins with common-sense knowledge.[1] Accordingly, two significant aspects of common sense are harmonizing new knowledge with the familiar store of common-sense knowledge, a unifying and consensus-seeking aspect of integration; and adhering to a standard of validity, integrating knowledge based on the norms and values of one's community. Later, the standards of validity become heightened as the community becomes more specialized and disciplinary.[2] Looking at the ways students considered multiple perspectives and their critical analysis in the term papers (a literature review), they were using common sense (per Gadamer) to engage with what was not familiar to them as a basis of that integration. For Boix-Mansilla and Gardner, common sense is the first step to going beyond disciplinary knowledge that involves drawing on one's own ideas to make sense of phenomena.[3] They argue that "much of what is termed 'interdisciplinary' work is actually predisciplinary work—that is, work based on common sense, not on the mastery and integration of a number of component disciplines."[4] Notwithstanding this, the students in Weird Science were able to demonstrate a truly interdisciplinary understanding, the ability to integrate multiple disciplines to support their answers to the complex question posed by the course regarding what it means to be human.

Two guest lecturers in this course, an economist and a sociologist, provided disciplinary perspectives for the students to integrate. The economist explored the enduring question of what it means to be human in the context of an economic system that seeks to condition and shape human economic behavior for the purpose of perpetuating the existence and survival of that same system. The sociologist questioned human insatiability. Both wrote case studies that introduced challenging concepts that asked students to consider alternative perspectives that facilitated the re-thinking of some widely held beliefs. These original case studies introduced the idea that behavior often explained as "human nature" is not a given, but instead a social construct.

Using the lens of constructivism and social constructivism proved to be an effective way to explore and interpret how students strive to become more information literate while conducting interdisciplinary research that is responsible. The information literacy and the responsible conduct of research modules emphasize skill sets to help students not only complete their assignments successfully; they also help to facilitate generalization to other contexts, which is the key to cultivating the ability to bring a variety of perspectives to bear on a problem that cannot be solved within a single discipline. Students were then able to use these skill sets to find quality information, to formulate the questions to ask when conducting animal or human subjects research, and to discover where to go for more information. When students are taken on ethical journeys informed by multiple disciplines, and they find that not only do these perspectives correspond with the course goals, but they also enhance the interdisciplinary habits of mind that promote lifelong learning.

These interdisciplinary habits of mind are crucial when considering complex problems, which have no disciplinary borders and must be solved using strategies that mimic their disciplinary un-restrictedness.[5] Paradoxically, finding solutions to the difficult, multi-pronged problems that we face will require innovative pedagogic initiatives that make simple common sense. As a result, educators should acknowledge that complex problems, and their solutions, often do not have disciplinary boundaries; understand that the disciplines involved in the problem before attempting to describe their interdisciplinary nature; promote interdisciplinary learning and its many advantages; introduce, expose, and engage students in interdisciplinary learning early and often; set clear, achievable, and measurable interdisciplinary student learning goals; and connect STEM and non-STEM approaches in interdisciplinary studies.[6]

INTERDISCIPLINARY LEARNING FROM A STUDENT'S PERSPECTIVE

This section discusses how interdisciplinary learning effectively promotes self-authorship, manifested through the learners' ability to reflect and base judgments on their knowledge,[7] as well as the ability to integrate multiple disciplines to accomplish a task. The learning partnership model of self-authorship assumes that knowledge is complex and socially constructed, that self is central to knowledge construction, and that authority and expertise are shared in the mutual construction of knowledge among peers.[8] Despite the variability and unpredictability of individual learner experience, student perspectives provide evidence for the unique challenges and distinct advantages of team-taught interdisciplinary studies. Below are the reflections of a former student, Tamrah D. Cunningham, who, as an undergraduate, participated in all of the interdisciplinary studies described in this book. She is currently a graduate student studying games design.

First-Year Learning Community

The "Story-Telling in Action-Adventure and Role-Playing Games" learning community[9] (LC) went through multiple iterations before its current version. The first iteration of this LC combined three courses, ENG 1101 (English Composition I), CST 1100 (Introduction to Computer Systems), and CST 1101 (Problem Solving with Computer Programming). As a student in this LC, I found, at first, the classes to be disjointed; in fact, I did not realize that I was a part of the LC until mid-semester. The final project, however, connected the courses by requiring the creation of a small, playable video game demo for the stories that students wrote. In addition, students had to write a research paper about some issue in games, such as gaming addiction. Because the three courses felt disjointed, it was difficult to make the interdisciplinary connections. However, the LC improved in subsequent iterations. The second iteration of the classes incorporated a game design document that students wrote in ENG 1101 and, instead of creating a video game demo in CST 1101, students were expected to create a video game trailer with interactive elements, and showcase it in both courses. In this iteration, however, we also introduced the idea of concept mapping.

Before this was implemented in the class, I was asked to create my own concept map for a role-playing game that I was working on at the time. I had never created a concept map before. I was more likely to write a simple outline of what I expected and what needed to be done and then do something completely different. However, through this exercise, I was able to get a better visual understanding of the usefulness of this process. I was able to see the relations of the elements of my game's narrative, see what did not make sense or was more entangled than I thought and go back and think to myself, "Okay, how do I unravel this mess of a story and make it to something that my players would see more clearly?" Having the students design their own concept map for their own narrative was a big help to them. Instead of just having a wall of text that students had to decipher every time they wanted to add something new to their story, they could have this visual tool that they could use that is easy to use, quick to make changes, and clearly see the connections between each plot relevant point that their game contained.[10]

Although this iteration worked better, it was not quite integrated because the CST 1100 course was still disconnected from the LC given the course's assignments (i.e., writing a research paper on artificial intelligence, focused on robotics rather than game mechanics) were not related to the other two classes. Thus, the LC was tweaked. To better incorporate the CST 1100 learning outcomes, the final paper was a game design document based on the game's stories that students wrote in ENG 1101, incorporating storyboard (screenshots), flowcharts, and pseudocode that were created in CST 1101. During this time, I had completed the LC and served as a Peer Advisor to students in the three courses. I observed that, in this iteration of the LC, students were able to really make the connection between computer programming and creative writing. By the end of the semester, students enjoyed the computer programming more because they were implementing their own stories, and this made the learning process meaningful and engaging.

While advising students, I saw my peers go through the same learning process that I went through. When I first enrolled in the three linked classes, a LC was something I had never experienced before. At first, I had no idea what the purpose of it was, or even that the assignments were related to each other. I never had taken classes where the subjects were interconnected, where what you did for one class was important to the other. The courses involved computer programming and technical writing, and, before I completed the semester, I had never considered the importance of writing in programming. I used to think that all I had to do to succeed in computer programming was to learn the coding language and type hundreds of lines of code in order to create a working program. However, it was nothing like that in the courses I took. The major difference was that I was not typing lines and lines of code in order to make a generic program. Instead, I was learning how to consider the problems that the English course presented; this made it possible to translate the video game narrative that I had written into an actual playable demo with the help of my class group mates. It was up to me to find solutions in programming in order to recreate the scenes of the story I wrote. Completing this project taught me how to solve computer programming problems creatively based on the limitations and scope of the story I had written. Also, due to this LC, many more opportunities opened up for me, including the chance to pursue my dream career as a game designer.[11]

Interpreting and Redefining Humanity from an Interdisciplinary Perspective

When asked via a survey about how they felt about taking a course like Weird Science, students revealed surprise at the type of class it turned out to be and the wealth of knowledge they ended up leaving the class with.[12] Even though they found it difficult to process the multiple perspectives at first, gradually, they were able to learn and make connections among the various disciplines. Students enjoyed the many guest lecturers who visited the class and were challenged by the various methods of framing the course's enduring question of what it means to be human.

The survey responses agree with my own experiences in the course. I had never considered really delving into the definition of what it means to be human. I accepted the definition given to me in my biology class and figured that, unless I was attempting to pursue a Philosophy major, there was no reason for me to further question this disciplinary perspective, that is, until I took Weird Science. As an interdisciplinary course, it introduced multiple fields of study, from Psychology to Mathematics, in order to answer the deceptively simple question of what it means to be human. I was interested to hear the differing points of view. For example, one perspective defined "human" as a being that possesses a soul and self-awareness, while another perspective disagreed, saying that what defines a human being is DNA and the ability to pass on genes so that the species can survive to a new generation. Due to the many opinions and viewpoints the class explored, I actually began to consider seriously what truly makes someone human. Do I go with the scientific definition, or do I go with a cultural belief that cannot be proven with hard facts? The opportunity to try to articulate my own definition was new to me.

During the class, I also undertook a research assignment wherein I created a concept map for all the various ideas that were presented in that class. I treated the concept map as a visual annotated bibliography that was extremely helpful for the term paper due at the end of the semester. Based on my thesis, I divided the concept map into portions based on the discipline, then divided the readings based on discipline and what the overall argument each author made based on the question of what it means to be human (e.g., the "nature" node connects to "Necessity vs. Desire" described in an excerpt focused on hunters and gatherers from the sociology guest lecturer's book.[13] That is, in the past, humanity took things only that were necessary to their survival. Today's desires are inflated by the consumerist nature that capitalism creates.).[14] By doing this, I was able to easily see where it was that I wanted to go with the paper and how to structure my arguments so everything flowed seamlessly.

Tying a definition of "human" to a virtual concept by predicting the future path of humanity in *Second Life*, a three-dimensional virtual world created by its residents, was a fun experience. I had to consolidate all that I heard from the variety of lectures, create my own opinion on what constitutes "being human," and then, based on my definition, make a prediction and design it in a virtual setting. It was interesting to hear the variety of definitions that my classmates had from the beginning, and then to see how, with the introduction of each new discipline, our definitions grew or changed in order to incorporate what we just learned.

Using Interdisciplinary Reasoning in Undergraduate Research

I worked on a few research projects in my time as an undergraduate student. These ranged from helping first-year students as a peer reviewer on an online student peer feedback program[15] to studying the effects of narrative-focused programming on creating my own role-playing game as part of the NSF-funded Louis Stokes Alliance for Minority Participation (LSAMP). I tried to branch out on what I was able to do during my four years working toward my Bachelor of Technology degree in Computer Systems. My most memorable research project was the New York City

Subway—Surface Air Flow Exchange (S-SAFE), a summer research opportunity in 2013. It was a collaboration between the Department of Energy, the Brookhaven National Laboratory, and the New York Police Department that was meant to highlight the importance of geoscience. The research was in two parts. The first part involved researching different geoscience topics and writing a paper, an educational brochure, and a presentation. My group and I had chosen the topic of earthquakes and spent the month researching how earthquakes are created, the devastation they can cause, and the ways that people can protect themselves during these natural disasters.

Writing a paper was something that many of us were not really comfortable with because many of us come from STEM-related fields. Though some disciplines required writing some form of research paper, these occasions were few and far in between, and most students just do not have the practice and honed writing skills. In order to help me more in this assignment, I decided to implement what I learned in my previous classes: concept mapping. Because I had made multiple concept maps before in different discipline, my mind would start to jump to looking at the question I was currently faced with and how it could be framed in a map. By looking at questions in this way, I was able to make the connections between ideas more quickly in order to determine the flow of writing. For a research paper, it was about the same thing. My question was, what are earthquakes and how do we clearly inform people about the dangers that these natural disaster can cause? I needed to identify the important information and determine how the information related to each other. I subconsciously mapped out the connections in my head and wrote the paper from there. Participating in this program helped me gain the skills I needed to approach something as daunting as a research paper, and produce a clear, informative, product. Subsequently, hearing the other undergraduate researchers' topics—which spanned from other natural disasters to human-made problems like fracking—was fascinating. The second part of the summer was spent collecting data around the city regarding the airflow of subway systems. We helped conduct what was known to be the largest airflow study in the USA; the findings were meant to increase our understanding of the risks that airborne contaminants pose in a large urban environment such as New York City. It was almost overwhelming to be a part of something so critical to the protection of millions of people, and it was a great experience overall.

Interdisciplinary Studies and General Education

Co-teaching CST 1102 Programming Narratives: Computer Animated Storytelling, an interdisciplinary course that stems from the LC is a unique experience because I am teaching the class that essentially started me on my career. However, this time, I am the one helping to impart to students the importance of such a course, as well as helping them to see the connections that can be made among reading classic works, writing (both creatively and technically), and computational thinking. It is my job to help them learn coding and show them that coding can be made easier and much more entertaining if they add a tale behind it, or in the case of this class, a video game story.

Students from several majors made great strides in this general education class. Most of the students had little to no understanding of coding, but, by the end of the semester, they were creating innovative scenes in *Alice*, a computer programming environment used to create three-dimensional animations, and really enjoying coding. Throughout the semester, the students were engaged in both the discussions about the various readings in the class and the problem-solving assignments. Students enjoyed debating each other in order to share their opinions and hear what their peers had to say. When it came to the programming aspect, students went above and beyond what was expected in this class; this is especially impressive considering that most of the students had no background in computer programming. They displayed a level of creativity that was a joy to witness. Many students approached the assignments with their own creative interpretations that they were glad to share with their peers. Many times, I witnessed students asking each other how they were able to make certain things happen, like a fiery fog effect in *Alice* or importing images created with industry-standard design applications. Students were eager to share what they discovered, and the entire process clearly demonstrated that they had learned how to solve problems collaboratively, using a *sensus communis* as a basis for the construction of new knowledge, just as Gadamer theorized.

Students were also taught how to create concept maps. Similar to the first-year learning community discussed above, students were able to map out their background story and the subsequent branching story paths (i.e., side quests) related to the story and create the connections between the various narrative branches that their story took. However, they also used the concept map for another purpose. In order to make sure that they had a firm grasp on the computational terms introduced throughout the course, students were asked to create a concept map that showed the connections between all the terms so that they could understand (e.g., algorithms, objects, classes, methods, and properties) and see that there is always an inherent relationship between these various ideas they are introduced to necessary for computational thinking.

There is, however, still work to be done to figure out how reading classic literature, creative and technical writing, and computer programming can be best integrated to create a video game prototype while helping students transfer skills between general education courses and specialized professional courses. These revisions are being incorporated in the current, second, iteration of this course. For example, early on in the course, students are tasked with reading and annotating Leo Tolstoy's short story, "The Three Questions," and then creating a scene from this story in *Alice* explaining its significance, as well as applying the hero's journey plot structure to *Oedipus the King* and providing ways this play could be translated into an action-adventure or role-playing game. This type of assignment marries the type of critical analysis, reflective thinking, creativity, and the construction of new knowledge that is at the heart of technology-supported interdisciplinary strategies.

PLACE-BASED LEARNING IN VIRTUAL WORLDS

Free educational technologies were used to scaffold student learning during the creation of the projects noted in the chapters, namely, *Alice*, *Second Life*, and the *Visual Understanding Environment*. While *Alice* provided a realistic setting for students to create and program their game world, *Second Life* provided a space for depicting their imaged future of humanity, and fostered place-based learning in virtual worlds.[16] The function of *Visual Understanding Environment*, on the other hand, was to help students conceptualize their ideas both in their written work and in their virtual worlds. A free application that provides a flexible visual environment for structuring, presenting, and sharing digital information, *Visual Understanding Environment* provided support for in-depth analysis of concept maps, with the ability to merge maps. Altogether, these technologies helped foster student collaborative problem-solving, creativity, critical thinking, and the creation of new knowledge.

NOTES

1. Hans-Georg Gadamer, *Truth and Method*, trans. Weinsheimer, Joel and Marshall, Donald G. (New York: Continuum, 1995. For a more detailed analysis of this idea, see Chap. 1 of the present book.
2. Ibid.
3. Veronica Boix-Mansilla and Howard Gardner, "Teaching for Understanding—Within and Across the Disciplines," *Educational Leadership* 51, no. 5 (1994): 14–18.
4. Ibid., 17.
5. Karl R. Popper, *Conjectures and Refutations: The Growth of Scientific Knowledge* (New York: Routledge and Kegan Paul, 1963).
6. See last chapter of the present book, which concludes with six recommendations. Also see Project Kaleidoscope, *What Works in Facilitating Interdisciplinary Learning in Science and Mathematics: Summary Report* (Washington, DC: AAC&U, 2011); Lisa R. Lattuca, Lois J. Voigt, and Kimberly Q. Fath, "Does Interdisciplinarity Promote Learning? Theoretical Support and Researchable Questions," *Review of Higher Education* 28, no. 1 (2004): 23–48.
7. Marcia B. Baxter Magolda, "Three Elements of Self-Authorship," *Journal of College Student Development* 49, no. 4 (2008): 269–284; Marcia B. Baxter Magolda, *Making Their Own Way: Narratives for Transforming Higher Education to Promote Self-Development* (Sterling, VA: Stylus, 2001).
8. Magolda, *Making Their Own Way*.
9. Learning communities encourage integration of learning across courses. See George D. Kuh, *High-Impact Educational Practices: What They Are, Who Has Access to Them, and Why They Matter* (Washington, DC: AAC&U, 2008).
10. See Reneta D. Lansiquot and Candido Cabo. "Concept Mapping Narratives to Promote CSCL and Interdisciplinary Studies," in *Proceedings of the Computer Supported Collaborative Learning (CSCL) Conference 2015*, vol. 2 (New York: ACM, 2015), 504–507.
11. See Reneta D. Lansiquot, Candido Cabo, and Tamrah D. Cunningham, "Playing between the Lines: Promoting Interdisciplinary Studies with Virtual Worlds," in *The Role-Playing Society: Essays on the Popular Influence of Role-Playing Games*, eds. Andrew Byers and Francesco Crocco (Jefferson, NC: McFarland, 2016); David R. Krathwohl, "A Revision of Bloom's Taxonomy: An Overview," *Theory into Practice* 41, no. 4 (2002): 143–162; Lev S. Vygotsky, *Mind in Society: The Development of Higher Psychological Processes* (Cambridge, MA: Harvard University Press, 1978).

12. Also see Marlene Hidalgo, "Interdisciplinary Learning from a Student's Perspective," in *Cases on Interdisciplinary Research Trends in Science, Technology, Engineering, and Mathematics: Studies on Urban Classrooms*, ed. Reneta D. Lansiquot (New York: Information Science Reference, 2013), 1–18.

13. Costas Panayotakis, *Remaking Scarcity: From Capitalist Inefficiency to Economic Democracy*, London: Pluto Press, 2011, 38–51. For additional information on his perspective, see Chap. 2 of the present book.

14. Figure 2 in Reneta D. Lansiquot "Towards Open-Source Virtual Worlds in Interdisciplinary Studies," in *Encyclopedia of Information Science and Technology* (3rd ed.), ed. Mehdi Khosrow-Pour (New York: Information Science Reference, 2015), 2650. Also see Lansiquot and Cabo, "Concept Mapping Narratives," 504–507.

15. For a description of this program, see Reneta Lansiquot and Christine Rosalia, "Online Peer Review: Encouraging Student Response and Development," *Journal of Interactive Learning Research* 26, no. 1 (2015): 105–123.

16. Reneta D. Lansiquot, Janet Liou-Mark, and Reginald A. Blake, "Learning Geosciences in Virtual Worlds: Engaging Students in Real-World Experiences," in *Education and Information Technology Annual 2015: A Selection of AACE Award Papers*, ed. Theo J. Bastiaens and Gary H. Marks (Chesapeake, VA: AACE, 2015), 223–240.

BIBLIOGRAPHY

Alice [software]. Accessed August 22, 2015. http://www.alice.org

Baxter Magolda, Marcia B. 2001. *Making Their Own Way: Narratives for Transforming Higher Education to Promote Self-Development*. Sterling, VA: Stylus.

——— 2008. "Three Elements of Self-Authorship." *Journal of College Student Development* 49 (4): 269–284.

Boix-Mansilla, Veronica, and Howard Gardner. 1994. "Teaching for Understanding—Within and Across the Disciplines." *Educational Leadership* 51 (5): 14–18.

Gadamer, Hans-Georg. 1995. *Truth and Method*. Translated by Joel Weinsheimer and Donald G. Marshall, NewYork: Continuum.

Hidalgo, Marlene. 2013. "Interdisciplinary Learning from a Student's Perspective." In *Cases on Interdisciplinary Research Trends in Science, Technology, Engineering, and Mathematics: Studies on Urban Classrooms*, edited by Reneta D. Lansiquot, 1–18. New York: Information Science Reference. doi: 10.4018/978-1-4666-2214-2.ch001

Krathwohl, David R. 2002. "A Revision of Bloom's Taxonomy: An Overview." *Theory into Practice* 41 (4): 212–218.

Kuh, George D. 2008. *High-Impact Educational Practices: What They Are, Who Has Access to Them, and Why They Matter.* Washington, DC: AAC&U.

Lansiquot, Reneta, and Christine Rosalia. 2015. "Online Peer Review: Encouraging Student Response and Development." *Journal of Interactive Learning Research* 26 (1): 105–123.

Lansiquot, Reneta D. 2015. "Towards Open-Source Virtual Worlds in Interdisciplinary Studies." In *Encyclopedia of Information Science and Technology*, edited by Mehdi Khosrow-Pour, 3rd ed, 2647–2653. New York: Information Science Reference.

Lansiquot, Reneta D., and Candido Cabo. 2015. "Concept Mapping Narratives to Promote CSCL and Interdisciplinary Studies." Proceedings of the Computer Supported Collaborative Learning (CSCL) conference 2015, 504–508. New York: ACM.

Lansiquot, Reneta D., Janet Liou-Mark, and Reginald A. Blake. 2015. "Learning Geosciences in Virtual Worlds: Engaging Students in Real-World Experiences." In *Education and Information Technology Annual 2015: A Selection of AACE Award Papers*, edited by Theo J. Bastiaens and Gary H. Marks, 223–240. Chesapeake, VA: AACE.

Lansiquot, Reneta D., Candido Cabo, and Tamrah D. Cunningham. 2016. "Playing between the Lines: Promoting Interdisciplinary Studies with Virtual Worlds." In *The Role-Playing Society: Essays on the Popular Influence of Role-Playing Games*, edited by Andrew Byers and Francesco Crocco, 143–162. Jefferson, NC: McFarland.

Lattuca, Lisa R., Lois J. Voigt, and Kimberly Q. Fath. 2004. "Does Interdisciplinarity Promote Learning? Theoretical Support and Researchable Questions." *Review of Higher Education* 28 (1): 23–48.

Linden Lab. *Second Life* [software]. Accessed August 22, 2015. http://www.secondlife.com

Panayotakis, Costas. 2011. *Remaking Scarcity: From Capitalist Inefficiency to Economic Democracy.* London: Pluto Press.

Popper, Karl R. 1963. *Conjectures and Refutations: The Growth of Scientific Knowledge.* New York: Routledge and Kegan Paul.

Project Kaleidoscope. 2011. *What Works in Facilitating Interdisciplinary Learning in Science and Mathematics: Summary Report.* Washington, DC: AAC&U.

Visual Understanding Environment [software]. Accessed August 22, 2015. http://vue.tufts.edu

Vygotsky, Lev S. 1978. *Mind in Society: The Development of Higher Psychological Processes.* Cambridge, MA: Harvard University Press.

INDEX

© The Editor(s) (if applicable) and The Author(s) 2016   141
R.D. Lansiquot (ed.), *Interdisciplinary Pedagogy for STEM*,
DOI 10.1057/978-1-137-56745-1

Printed in the United States
By Bookmasters